AF517841

ENCYCLOPAEDIA OF ENGINEERING

ENCYCLOPAEDIA OF ENGINEERING

VOL. 2

CIVIL ENGINEERING

E-P

By

A.M. Kharmawphlang
A. Khongphai

AKANSHA PUBLISHING HOUSE
NEW DELHI

AKANSHA PUBLISHING HOUSE
4649-B/21, Ansari Road
Darya Ganj, New Delhi-110 002
Email: ektabooks@yahoo.com
Ph.: 23263193/9811582579
Fax: 011-23263193

Encyclopaedia of Engineering

First Published 2006

ISBN 81-8370-048-9 (Set)

[The responsibility for the facts stated, conclusions reached etc. is entirely that of the authors. The publisher is not responsible for them, whatsoever.]

PRINTED IN INDIA

Published by M.P. Misra for Akansha Publishing House, New Delhi and Printed at Aryan Enterprise, Delhi.

PREFACE

Engineering is the application of science to the needs of humanity. This is accomplished through the application of knowledge, mathematics, and practical experience to the design of useful objects or processes.

Engineering involves the knowledge of mathematical and natural sciences (biological and physical) gained by study, experience, and practice, applied with judgement and creativity to develop ways to utilize the materials and forces of nature for the benefit of mankind.

Encyclopaedia of engineering contains engineering terms with clear and concise definitions. The encyclopaedia is highly illustrated and written in plain, non-technical language and extensive cross-referencing makes concepts more easily understood. Timings and Twigg focus on a wide range of engineering disciplines — engineering science, electrical and electronic engineering, workshop practices and mechanical engineering — and provide definitions that are relevant to all.

This vital reference offers a wealth of essential information in a convenient, quick-find format. Whether you're a professional, a student, a writer, or a general reader with an interest in science, there is no better or more authoritative way to stay up-to-speed with the current language of the various branches of engineering or gain an understanding of their key ideas and concepts.

Encyclopaedia of Engineering is written specifically for further education students by experienced authors who have taught and written extensively at this level.

E

E&O Insurance. Errors and Omissions Insurance.

E. 1. Represents the modulus of elasticity. 2. Represents volts (electromotive force).

EA. Exhaust Air.

Eagle Beak. A 6-inch x 3/4 inch outside çorner ceramin tile trim shape tile.

Earlywood. Springwood.

Earth Pigment. Pigments which occur as deposits in earth and are removed by mining; such pigments as a whole are permanent in color, non-bleeding, and are not readily changed by heat, light, moisture, and alkalis.

Earthquake. Movement in the earth's crust that produces horizontal and vertical movement on the ground surface.

Earthwork. An embankment or other construction made of earth; any work involving movement or use of soil and other earthen material.

Eased Edges. 1. The slight rounding of the corners of a piece of finish lumber. 2. A tapered, slightly rounded factory edge of gypsum board.

Easement. 1. A right to utilize real property owned by another. 2. An interest in land owned by another that entitles its holder to a specific limited use.

Eastern Frame. See Balloon Frame.

Eastern Method. See Pick and Dip.

EAVES. The part of a roof that extends beyond the exterior walls below it.

Eave Trough. Gutter for catching rain water at the eaves of a sloping roof.

Eave Vent. A usually screened opening at the eave line to allow a free flow of ventilation air into the underroof area.

Eccentric. Not having common centers; offcenter.

Eccentrically Braced Frame (EBF). A diagonal braced frame in which a least one end of each bracing member connects to a beam a short distance from a beam-to-column connection or from another beam-to-brace connection.

Eccentricity. The distance from the application of a structural load to the axis or centroid of the carrying member.

Eclectic. Architectural design based on picking and choosing from various styles; see Pastiche.

Ecology. A branch of science concerned with the interrelationship of organisms and their environments.

Economic. Profitable; careful, efficient, and prudent use of resources.

Economic Life. The period of time a property will produce sufficient income after expenses to justify its continued operation.

Economic Obsolescence. Depreciation; the reduction in value of real property caused by extrinsic factors such as a declining neighborhood, a factory closure, or unsuitable zoning.

Economizer Cycle. Not an individual system, it is a modification of the single zone system, terminal reheat, multi-zone, dual duct, and variable air volume (VAV); it is a modification that controls and adjusts motorized dampers to draw in outside air when it is advantageous for reduced cooling energy.

Economy. Efficient and concise use of resources.

Economy Brick. A brick larger than standard.

Ecosystem. A biological community of interacting organisms and their physical environment.

Eco-tech. Green design.

Edge. 1. The cutting side of a blade. 2. The degree of sharpness of a tool. 3. The line where an object or area begins or ends. 4. The extreme verge or brink of a cliff. 5. Of gypsum board, the paper bound edge as manufactured.

Edge Bead. A strip of metal or plastic used to make a neat, durable edge where plaster or gypsum board abuts another material.

Edge Distance. The distance from a rivet, bolt, screw, or nail to the edge of a structural member.

Edge, Drip. See Drip Edge.

Edge Form. A forming member used to limit the horizontal spread of fresh concrete on flat surfaces.

Edge-Grained Lumber. Lumber that has been sawed so that the wide surfaces extend approximately at right angles to the annual growth rings. Lumber is considered edge grained when the rings form an angle of 45 degrees to 90 degrees with the wide surface of the piece.

Edge Joint Cracks. In concrete or asphalt road paving, the separation of the joint between the pavement and the shoulder, commonly caused by the alternate wetting and drying beneath the shoulder surface; other causes are shoulder settlement, mix shrinkage, and trucks straddling the joint.

Edge Joint. The place where two pieces of wood are joined together edge to edge, commonly by gluing; the joint may be made by gluing two squared edges as in a plain edge joint or by using machined joints of various configuration, such as tongued-and-grooved joints.

Edge-Matched. 1. See Tongue and Groove. 2. Lumber that has been rabbeted on both edges of each piece; in either case, the purpose is to provide a close joint when fitting two pieces together.

Edge Mounted Tile. A type of mounted tile wherein tile is assembled into units or sheets and are bonded to each other at the edges or corners of the back of the tiles by an elastomeric or resinous material which becomes an integral part of the tile installation.

Edge Preparation. The prepared shape on the edge of metal for welding.

Edger. A finishing tool used on the edges of fresh concrete to provide a rounded corner.

Edge Repair. The repair of an edge of a construction member by the use of plaster or concrete.

Edge Sheets. Felt strips that are cut to widths narrower than the standard width of the full felt roll; used to start the felt-shingling pattern at a roof edge.

Edge Stripping. Application of felt strips cut to narrower widths than the normal 36 inch width of a felt roll, used to start the felt-shingling pattern at a roof edge.

Edge Venting. The practice of providing regular spaced protected openings around a roof's perimeter to relieve the water vapor.

Edging. Striping.

Edifice. A large building.

EDM. Electronic Distance Measuring.

EE. Electrical Engineer.

Effective. 1. Capable of producing a decided, decisive, or desired effect. 2. Capable of performing the particular function specified with safety.

Effective Area. Actual flow area of an air inlet or outlet; gross area minus area of vanes or grille bars.

Effective Area of Concrete. The area of a section which lies between the centroid of the tension reinforcement and the compression face of the flexural member.

Effective Area of Reinforcement. The area obtained by multiplying the cross sectional area of the reinforcement by the cosine of the angle between its direction and the direction for which the effectiveness is to be determined.

Effective Depth of Section. The distance from the extreme compression fiber to the centroid of tension reinforcement.

Effective Opening. The minimum cross-section area of the opening where water is discharged from a water supply pipe.

Effective Prestress. The stress remaining in prestressing tendons after all losses have occurred, excluding effects of dead load and superimposed load.

Effective Stiffness. The value of the lateral force in the isolation system, or an element thereof, divided by the corresponding lateral displacement.

Effective Temperature. Overall effect on a human of air temperature, humidity, and air movement.

Efficiency. The ratio of useful work performed to the total energy expended or heat taken in.

Efficiency Living Unit. Any room having access to bathroom facilities and having cooking facilities and intended or designed to be used for combined living, dining, and sleeping purposes.

Efflorescence. A deposit of water soluble salts on the surface of masonry or plaster caused by the dissolving of salts present in the masonry; migration of the solution to the surface and deposition of the salts when the water evaporates; the surest preventative of efflorescence is to keep water out of masonry; the process of efflorescing is also referred to as Whiskering or Saltpetering.

Effluent. Fluid discharge from a sewage system.

Egg and Dart. A decorative carved molding consisting of alternating egg-shaped and arrow-shaped parts.

Egg Shell. Finish that closely resembles the luster of an egg shell.

Eggshelling. 1. The texture of a fired glaze similar in appearance to the surface of an eggshell. 2. Chip-cracked plaster, either base or finish coat; the form taken is concave to the surface and the bond is partially destroyed.

Egress. The way out; exit.

EIA. 1. Environmental Impact Assessment. 2. Electronic Industries Association.

EIFS. Exterior Insulation Finish System.

Eight Bend. A pipe fitting which allows the run of pipe to make a 45 degree bend.

EIP. Membrane is generally reinforced with polyester fabric. Resists fire, chemicals, oils and tears.

EIR. Environmental Impact Report.

EIS. Environmental Impact Statement.

Ejector. Device which uses high fluid velocity, such as a venturi, to create low pressure or vacuum at its throat to draw in fluid from another source.

Elapsed Time Indicator. A mechanical or electronic device which measures the passage of time.

Elastic. Able to return to its original size and shape after removal of stress.

Elastic Analysis. An analysis of deformations and internal forces based on equilibrium, compatibility of strains, and assumed elastic behavior and representing to suitable approximation the three dimensional action of the structural shell together with its auxiliary members.

Elastic Limit. The largest stress which a material can withstand without being permanently deformed.

Elastic Sheet Roof. A thin, pliable roofing material that is able to expand and contract to its original size regardless of weather conditions.

Elastic Shortening. In prestressed concrete, the shortening of a member which occurs on the application of forces induced by prestressing.

Elastomer. Any of various elastic substances resembling rubber; a material which at room temperature can be stretched repeatedly to at least twice its original length and, upon release of the stress, will return with force to its approximate original length.

Elastomeric. Having the characteristics of an elastomer.

Elastomeric Flashing. A rubber-like material used as flashing on a roof system.

Elastomeric/Plastomeric Membrane. A rubber-like sheet material used as a roof covering.

Elbow. A pipe fitting having two openings which allows a run of pipe to change directions.

Elbow Catch. A spring-loaded device consisting of a rocker arm and angle strike, for locking the inactive leaf of a pair of cabinet doors.

Electrical. Relating to, or operated by electric current.

Electrical Conductivity. The conducting power of a specified material; the reciprocal of electrical resistivity.

Electrical Curing. A system in which a favorable temperature is maintained in freshly-placed concrete by supplying heat generated by electrical resistance.

Electrical Engineer. An engineer who designs electrical systems.

Electrical Fee. The amount of money charged for the inspection or installation for the electrical wiring work in a building or structure.

Electrically Welded Wire Fabric. Large dimension wire mesh used for reinforcing concrete slabs on grade. Same as Electrically Welded Wire Mesh.

Electrically Welded Wire Mesh. Electrically Welded Wire Fabric.

Electrical Metallic Tubing. Unthreaded light weight piping for running electrical conductors; easier to handle than rigid conduit and installed more rapidly because of the type of non-threaded fittings used with it; also called Thin Wall Conduit.

Electrical Porcelain. Vitrified whiteware having an electrical insulating function.

Electrical Potential. Electrical force which moves, or attempts to move, electrons along a conductor or resistance; measured in volts.

Electrical Resistance (R). The difficulty electrons have moving through a conductor or substance.

Electrical Resistivity. The resisting quality of a specified material; the reciprocal of electrical conductivity.

Electrical Socket. A receptacle for connecting electrical appliances to the electrical supply.

Electric Boiler. A tank in which water is heated or hot water is stored, controlled by an electric current.

Electric Cord. A small flexible insulated electrical cable having a plug at one or both ends used to connect a power tool or mechanical device with a receptacle which supplies electric current.

Electric Defrosting. Use of electric resistance heating coils to melt ice and frost off evaporators during defrosting.

Electric Elevator. See Electric Traction Elevator.

Electric Eye. 1. Photoelectric cell; a miniature cathode ray tube.

Electric Field. A region of electrical influence.

Electric Furnace. An enclosed structure in which heat is produced, controlled by electric current.

Electric Grinder. A mechanical device powered by electric current that wears down, polishes, or sharpens by friction.

Electric Heater. A heat-producing unit powered by electricity.

Electric Heating. System in which heat from electrical resistance units is used to heat the building.

Electric Hoist. An apparatus for lifting people or materials powered by electric current.

Electrician. A craft worker who installs, maintains, and repairs electrical systems in buildings.

Electric Insulation. Substance which has almost no free electrons.

Electric Lockset. A doorknob and deadbolt assembly in a door locked or unlocked by electric current.

Electric Manhole. An access hole for the service of underground electric lines; also used for pull stations when laying electric cable.

Electric Meter. An instrument for measuring consumption of electrical power.

Electric Pole. A vertical wooden pole used to carry electric utility wires. Opposite charged terminals, as in an electric cell or battery.

Electric Range. A cooking stove with an oven and a flat top with burners, powered by electric current.

Electric Sign. A fixed or portable, self-contained electrically-illuminated appliance with words or symbols designed to convey information or attract attention.

Electric Steamer. A device which produces steam to strip old wallpaper off of wall surfaces.

Electric Strike. An electric device that permits releasing of the door latch from a remote control.

Electric Traction Elevator. Elevator operated by electric motor and suspended from cables.

Electric Water Valve. Electrically operated solenoid valve controlling water flow.

Electrochemical Coating. A coating on metal by means of electron transfer by electrical current; electroplating; electrodeposition.

Electrode. Terminal point to which electricity is brought in the welding operation and from which the arc is produced to do the welding; in electric arc welding, the electrode is usually melted and becomes part of the weld.

Electrode Force. In spot, seam, and projection welding, the force, in pounds, between the electrodes during the actual welding cycle.

Electrode Holder. A device used for mechanically holding the electrode and conducting current to it.

Electrolier. Pole-mounted street light.

Electrolysis. Production of chemical changes, such as decomposition, by the passage of current through an electrolyte, an acidic liquid, or damp earth; corrodes metals.

Electrolytic Condenser-Capacitor. Plate or surface capable of storing small electrical charges.

Electromagnet. A soft metal core made into a magnet by the passage of electric current through a coil surrounding it.

Electromotive Force. The force that makes electrons move in an electric current; voltage.

Electron. Elementary particle or portion of an atom which carries a negative charge.

Electronic. Pertaining to electrons.

Electronic Air Filter. A filter that attracts dust particles by static charge; also called Electrostatic Filter.

Electronic Distance Measuring. A high-precision surveyor's instrument that measures distance by use of radio-frequency or light-frequency electromagnetic waves that are reflected back to source, the elapsed time being precisely measured and converted to distance.

Electronic Leak Detector. Electronic instrument which measures electronic flow across gas gap; electronic flow changes indicate presence of refrigerant gas molecules.

Electronics. A branch of physics and technology concerned with the behavior and movement of electrons in a vacuum, gas, or semiconductor and with electronic devices.

Electroplate. Coat by electrolytic deposition with chromium, silver, copper, or other metal.

Electrostatic Coating. Painting with a spray that utilizes electrically charged particles to ensure complete coverage.

Electrostatic Filter. See Electronic Air Filter.

Element. 1. A component part. 2. Any of the hundred or so substances that cannot be resolved by chemical means into simpler substances.

Elevated Floor. A floor system not supported by a subgrade.

Elevated Slab. A roof slab or floor supported by structural members.

Elevated Slab, Concrete. A concrete roof system or concrete flooring system supported by structural members.

Elevated Slab Formwork. The system of support for a freshly poured or placed concrete elevated slab.

Elevated Slab Reinforcing. Metal or steel bars embedded in freshly poured concrete to strengthen an elevated slab.

Elevated Stairs. A stair system not supported by the subgrade.

Elevated Temperature Testing. Tests on plastic pipe above 23° C.

Elevation. 1. The height above a given level, such as sea level. 2. High place or position. 3. A drawing or diagram made by projection on a vertical plane; a flat drawing of the front, side, or back of a building.

Elevator. 1. A hoisting machine in a shaft;a cable or chain hoist conveying system used for raising material or passengers in a cab, cage, or platform. 2. A building for elevating, storing, discharging, and sometimes processing grain; a grain elevator.

Elevator Penthouse. A roof structure containing the machine room for an electric traction elevator.

Elevator Shaft. A lined vertical shaft in a building, usually of a fire-rated construction, in which the elevator cab ascends and descends.

Ellipse. 1. A closed plane curve generated by a point moving in such a way that the sums of its distances from two fixed points is a constant. 2. An oval.

Ellipsoid. A solid like a flattened sphere.

Elliptical. Shaped like an ellipse.

Elongation. The state of being lengthened.

Embankment. A fill whose top is higher than the adjoining surface.

Embed. In application of gypsum wallboard, to apply and cover joint tape with joint compound.

Embedment. 1. The process of pressing a felt, aggregate, fabric, mat, or panel uniformly and completely into hot bitumen or adhesive. 2. The process of placing a material into another material so that it becomes an integral part of the whole material.

Embedment Length. The length of embedded steel reinforcement provided beyond a critical section.

Embezzlement. Theft of property which has become a possession of the thief; to appropriate property, that is entrusted to one's care, fraudulently for one's own use.

Emboss. 1. Carve or mold in relief. 2. Form designs so that they stand out on a surface. 3. Make protuberant. 4. Adorn or embellish.

Embossed. In carpet, the type of pattern formed when heavy twisted tufts are used in a ground of straight yarns to create an engraved appearance; both the straight and twisted yarns are often of the same color.

Embossed Paper. Wallpaper run through rollers with raised areas to provide a light relief effect.

Emergency Generator. A gasoline powered motor and electrical generator provided for emergency lighting and power during interruption of the normal electrical supply.

Emery. Slow-cutting, short-lived abrasive.

EMF. Electromotive Force.

Eminent Domain. The power of the state to take private property for public use, upon payment of a fair price.

Emittance. A measure of the propensity of a material to give off thermal radiation.

Empirical Formula. A formula that is developed from experience rather than from scientific theory.

Empty-Cell Process. A method of obtaining deep penetration of a wood preservative with a relatively low net retention of the preservative; surplus preservative is removed from the wood cells by vacuum.

EMT Conduit. Electrical Metallic Tubing.

Emulsified Asphalt. An emulsion of asphalt cement and water that contains a small amount of an emulsifying agent, a heterogeneous system containing two normally immiscible phases (asphalt and water) in which the water forms the continuous phase of the emulsion, and minute globules of asphalt form the discontinuous phase. Emulsified asphalt may be of either the anionic, electronegatively charged asphalt globules, or cationic, the electropositively charged asphalt globule types, depending upon the emulsifying agent.

Emulsified Asphalt Mix (Cold Mix). A mixture of emulsified asphalt and aggregate; produced in a central plant (plant mix) or mixed at the road site (mixed-in-place).

Emulsifier. Material which, when added to a mixture of dissimilar materials will produce a stable emulsion.

Emulsifying Agents. Substances of chemical nature that intimately mix and disperse dissimilar materials ordinarily immiscible, such as oil and water, to produce a stable emulsion; a substance which when added to a liquid permits suspension of fine particles or globules in the liquid.

Emulsion. 1. A fine dispersion of one liquid in another, as in paint; emulsion paint is a water-thinned paint containing a non-volatile substance as its binding medium. 2. In roofing, a coating consisting of asphalt and fillers suspended in water.

Emulsion Paint. Water-thinned paint with an emulsified oil, resin, or latex vehicle.

Enamel. Type of paint made by grinding or mixing pigments with varnishes or lacquers.

Enameling. The process of painting with enamel.

Enamel Paint. A paint which dries to a hard gloss or semi-gloss smooth finish.

Encased Burial Conduit. Metal or plastic conduit EB for outdoor wiring with type TW wires encased, or type UF cable.

Encased Knot. A knot whose rings of annual growth are not intergrown with those of the surrounding wood.

Enclosed. Surrounded by a case which will prevent a person from accidentally contacting wiring, equipment or live parts contained therein.

Enclosed Switch. A electric switch which is protected by thin metal shields on either side of the switch to prevent accidental tripping of the device.

Enclosure. The case or housing of electrical or mechanical apparatus, or the fence or walls surrounding an installation to prevent personnel from accidentally contacting energized parts, or to protect the equipment from physical damage.

Enclosure, Telephone. Partitions which provide privacy on two sides of a public telephone, or which completely surround the user; a phone booth.

Encroachment. Personal property of one person intruding upon real estate owned by another.

Encumbrance. A charge against real property.

End. 1. The extreme point of an object. 2. The termination of an operation. 3. Of gypsumboard, the end perpendicular to the paper- bound edge; the gypsum core is always exposed.

End-Bearing Pile. A pile calculated to carry all of its load on its point, neglecting skin friction.

End Bell. 1. Cast iron pipe with a wide opening at one end, to receive the small end of an adjoining pipe; see Bell and Spigot Joint. 2. End structure or plate of electric motor which usually holds motor bearings.

End Burn. Over-calcined gypsum board resulting in easily damaged, soft, fragile ends.

End Distance. The distance from a bolt, screw, or nail to the end of a wood structural member.

End-Grained Wood. The grain as seen on a cut made at a right angle to the direction of the fibers.

End Joint. A joint made by bonding two pieces of wood together end to end, commonly by finger or skarf joint or butt joint.

End Lap. The overlapping of roofing felts at the ends of sheets cut off the roll.

End Matched. A board with a tongue and groove joint on the ends as well as on the sides.

End Nail. To drive a nail through one piece of lumber and into the end grain of another.

Endothermal. Chemical reaction in which heat is absorbed.

Endothermic Reaction. A process that requires and absorbs heat.

End Play. Slight movement of a shaft along its center line.

End Support. Bearing point for a metal open-web joist located at the end of the joist, usually constructed of a steel plate attached to a supporting component to uniformly distribute the load to a supporting component.

End Truss. A factory made wooden truss which is used at the gable end of a building and to which the sheathing and siding are fastened.

End Wrench. A hand tool with one or both ends shaped to grip a nut or bolt head that is to be held or turned.

Energy. Usable power such as heat or electricity; actual or potential ability to do work.

Energy Absorption. Energy is absorbed as a structure distorts.

Energy Conservation. Measures taken to reduce if not eliminate unnecessary use of all forms of energy and the consumption of non-renewable resources.

Energy Dissipation. Reduction in intensity of earthquake shock waves with time and distance.

Energy Efficient Standards. Building code, which sets standards for energy conservation in buildings and facilities to which the code applies.

Enforcing Agency. The designated department or agency of any city, county or the state as specified in the statutes.

Engaged Column. A building column that is partially subsumed into the wall.

Engineer. A person who is professionally qualified in a branch of engineering.

Engineered Fill. Earth compacted into place in such a way that it has predictable physical properties, based on laboratory tests and specified, supervised, installation procedures.

Engineering. The application of science and mathematics by which the properties of matter and the sources of energy in nature are made useful to human beings in structures, machines, products, systems, and processes.

Engineering Fee. The amount charged for engineering services.

Engineer's Level. An accurate telescopic transit for measuring horizontal and vertical angles; used in ordinary surveying work; see Dumpy Level.

English Bond. Brickwork laid with alternating courses, each consisting entirely of headers or stretchers; also called Old English Bond or Dutch Bond.

English Chalk. Chalk obtained from the cliffs of England.

Engraved Plaque. A commemorative or identifying tablet where figures or letters are inscribed.

ENR. Engineering News Record; a construction industry publication.

Enrichments. Any cast ornament which cannot be executed by a running mould.

Entablature. In classical architecture, the horizontal parts just above the columns, consisting of the cornice, frieze, and architrave.

Entasis. A convex curvature added to the taper of the shaft of a column in the Greek and Roman styles as a design refinement to make the columns appear straight; also applies to similar adjustments to high walls and spires.

Enthalpy. A measure of the energy content of a system per unit mass; the sum of sensible and latent heat of a material.

Entrained Air. Microscopic air bubbles intentionally incorporated into mortar or concrete during mixing, usually by use of a surface-active agent, typically between 10 and 1,000 um in diameter.

Entraining Agent. A substance added to concrete, mortar, or cement that produces air bubbles during mixing, making it easier to work with and increasing its resistance to frost and freezing.

Entrance. The way in to a property, a building, or a room.

Entrance Door. The door which provides access to a building or structure.

Entrance Mat. A woven fabric or ribbed rubberized mat placed on the outside of a door threshold for the wiping of shoes.

Entrapped Air. Air in a concrete mix that enters from the atmosphere during mixing; after the concrete hardens and the excess moisture has evaporated, irregular holes remain; entrapped air bubbles are normally much larger and more irregular than entrained air bubbles.

Entrepreneur. One who assumes the risk of starting and operating a business.

Entropy. 1.A measure of the disorder or randomness of a system. 2. Mathematical factor used in engineering calculations. 3. Energy in a system.

Entry. 1. An entrance room. 2. The recording of a business transaction in a record.

Entry Lock. A deadbolt assembly mounted in a door which when unlocked allows access to a building or structure.

Entry Lockset. A combination doorknob and deadbolt assembly mounted in a door that when unlocked allows access to a building or structure.

Envelope. In roofing, a continuous edge that is formed by folding an edge of the base felt over the plies above and securing the base felt to the top felt, or if above-deck insulation is used, to the top surface of the insulation; the envelope thus formed will prevent bitumen dripping through the exposed edge joints of the laminated roofing membrane and also prevent water seeping into the insulation.

Environment. Physical surroundings, circumstances, and conditions that affect peoples' lives.

Environmental Factors. Conditions other than indoor air contaminants that cause stress, comfort, or health problems, for example, humidity extremes, drafts, lack of air circulation, noise, and overcrowding.

Environmental Impact Assessment. See Environmental Impact Report.

Environmental Impact Report. A report that examines and considers all likely effects on the environment caused by a proposed development of land, and proposes measures avoiding, reducing, or offsetting any significant adverse effects; also called Environmental Impact Assessment or Environmental Impact Statement.

Environmental Impact Statement. See Environmental Impact Report.

Environmental Sustainability. Land development and construction utilizing materials that are replaceable, like wood, and minimizing use of non-replaceable materials and energy resources.

Enzyme. Complex organic substance, originating from living cells, that speeds up chemical changes in foods; enzyme action is slowed by cooling.

Eolian. Borne, eroded, or deposited by the wind.

EPA. Environmental Protection Agency.

EPDM. Ethylene Propylene Diene Monomer.

Epicenter. The point on the earth's surface vertically above the origin, focus, or hypocenter of an earthquake.

Episcopal Throne. The bishop's ceremonial chair in a cathedral.

Epoxy Adduct. Epoxy resin having all of the required amine incorporated but requiring additional epoxy resin for curing.

Epoxy Adhesive. A two-part adhesive system employing epoxy resin and epoxy hardener used for bonding of ceramic tile to back-up materials.

Epoxy Amine. Amine-cured epoxy resin.

Epoxy Concrete. Concrete with added adhesive resin to aid in binding.

Epoxy Ester. Epoxy-modified oil; single package epoxy.

Epoxy Flooring. A resin spread on flooring to make a hard, tough surface.

Epoxy Grout. A two-part grout system consisting of epoxy resin and epoxy hardener, especially formulated to have impervious qualities, stain, and chemical resistance, used to fill joints between tile units.

Epoxy Mortar. A two-part mortar system consisting of epoxy resin and epoxy hardener used to bond tile to back-up material where chemical resistance and high bond strength is a consideration.

Epoxy Paint. Paint with added resin to give it adhesive and improved bonding characteristics in certain applications.

Epoxy Resin. An epoxy composition used as a chemical-resistant setting adhesive or chemical-resistant grout.

Epoxy Terrazzo. A two-part adhesive, employing epoxy resins, an epoxy hardener used for bonding marble or other stone chips set in portland cement to a backup material.

E.P.S. Expanded Polystyrene.

Equalizer. One that equalizes by distributing evenly or uniformly, as of force.

Equation. 1. A chemical equation shows what happens in a chemical reaction; the reactants equal the products. 2. A mathematical equation is a statement that two quantities or expressions are equal.

Equilateral Triangle. A triangle having three equal sides.

Equilibrium. A state of rest due to balanced forces; state of being in balance; implies no tendency to change.

Equilibrium Moisture Content. The moisture content at which wood neither gains or loses moisture when surrounded by air at a given relative humidity and temperature.

Equipment. The implements used in an operation or activity.

Equipment, Architectural. See Architectural.

Equipment, Insulation. See Insulation Equipment.

Equipment Mobilization. The assembly and movement of equipment to a jobsite.

Equipment Pad. A thick slab-type stone or precast concrete block placed under mechanical equipment to spread the weight and load of the machinery evenly and to prevent excessive vibration.

Equipment Rack. A device mounted on a wall to hold or store the implements used in an operation or activity.

Equitable Lien. A lien that is given to an improver of property to prevent the unjust enrichment of a property owner or a construction lender.

Equity. 1. The value of a firm's assets in excess of its liabilities. 2. The prevention of injustice that might result from strict enforcement of law.

Equivalent Static Force Analysis. A method by which a dynamic force is translated into an equivalent static force that produces a similar effect.

Equi-Viscous Temperature. The critical temperature at which asphalt reaches the optimum state of viscosity for adhesion to roofing felt.

Eraser. A rubber-like substance used by drafters to remove pencil or ink marks on drawings.

Erasing Machine. A small hand-held electric motor that rotates an eraser, used by drafters.

Erasing Shield. A thin metal stencil with various sizes and shapes of holes to enable a drafter to erase with precision.

Erect. 1. To build or construct by fitting parts together. 2. Vertical.

Erection. A building or structure.

Erg. A unit of work or energy, equal to the work done by a force of one dyne when its point of application moves one centimeter in the direction of action of the force.

Ergonomics. Applied science that investigates the impact of people's physical environment on their health and comfort; for example, determining the proper chair height for computer operators.

Erode. Gradual wearing away or destruction.

Erosion. 1. The gradual wearing away of material as a result of abrasive action. 2. Wear caused by moving water or wind. 3. Wearing away of paint films; heavy chalking tends to accelerate erosion.

Errors and Omissions Insurance. Insurance carried by architects and engineers to indemnify their clients against losses caused by the professional negligence of the architect or engineer; also called E and O Insurance, Professional Indemnity Insurance, or Malpractice Insurance.

Escalation. Increase in building cost due to inflation or other factors.

Escalator. A power-driven set of stairs arranged like an endless belt that ascend or descend continuously.

Escheat. The reversion of property to the state when a person dies without heirs and without a will.

Escrow. A neutral facilitator who follows the instructions of parties to a transaction.

Escutcheon. The protective, sometimes ornamental, plate around a door lock or pipe.

Essential Facilities. Structures or buildings which have been determined essential, and are intended to be safe and usable after an earthquake; such facilities include, but are not limited to. hospitals and other medical facilities having surgery or emergency treatment areas, fire and police stations, and government disaster and communication centers.

Essential Oils. Oils which have an odor, such as cedar oil or camphor oil.

Essential Services Act. An act that provides that essential services buildings shall be capable of providing essential services to the public after a disaster, shall be designed and constructed to minimize fire hazards, and to resist, insofar as possible, the forces generated by earthquake, gravity, and winds.

Estate. 1. The degree, quality, nature, or extent of one's ownership in property. 2. The assets of a deceased person. 3. Landed property, usually with a large house on it.

Ester. Organic compound formed from an alcohol and an organic acid by eliminating water.

Ester Gum. Resin produced synthetically by rosin reacting with glycerine.

Estimate. 1. A prediction of the cost of performing work; compute; calculate cost of a job. 2. A value judgement based on experience. 3. An approximation of construction costs.

Estoppel. The doctrine that a person may not contradict one's own positive representations.

E. Switch (Snap, General Use). A form of general use switch so constructed that it can be installed in flush device boxes or on outlet box covers, or otherwise used in conjunction with wiring systems recognized by these regulations.

ET. See Net Effective Temperature.

Etch. The art of producing designs on metal or glass by the use of the corrosive action of an acid; the use of acid to cut lines into metal or remove the surface of concrete.

Etch Acid. A chemical agent used in etching.

Etched Nails. Chemically treated nails to improve their holding power in wood framing.

Ethane. Low temperature application refrigerant.

Ethanol. Alcohol.

Ethyl Acetate. Rapid evaporating solvent made from ethyl alcohol and acetic acid.

Ethyl Alcohol. Alcohol produced by the distillation of fermented grain.

Ethyl Chloride. Toxic refrigerant now seldom used.

Ethyl Lactate. A solvent made by a reaction between ethyl alcohol and lactic acid.

Ethylene Plastic. Plastics based on resins made by the polymerization of ethylene or copolymerization of ethylene with one or more other unsaturated compounds.

Ethylene Propylene Diene Monomer (EPDM). A single-ply membrane of synthetic rubber, usually black or white, highly resistant to damage from ozone, ultraviolet radiation, weathering and abrasion; resists contamination from acids, alkalis, animal and vegetable oils, and oxygenated solvents such as ketones, esters, and alcohols; can be loosely laid, ballasted, or fully adhered.

ETS. Environmental Tobacco Smoke.

Eutectic. That certain mixture of two substances providing lowest melting temperature of all the various mixes of the two substances.

Eutectic Deformation. The composition within a system of two or more components which, on heating under specific conditions, develops sufficient liquid to cause deformation at minimum temperature.

Eutectic Point. Freezing temperature for eutectic solutions.

Eutectic Salts. 1. A group of materials that melt at low temperatures, absorbing large quantities of heat and then, as they recrystallize, release that heat. 2. One method used for storing solar energy.

Evacuation. Removal of air, gas, and moisture from a refrigeration or air conditioning system.

Evaporate. To pass off in vapor; to change a liquid into vapor or gas.

Evaporation. 1. The change of a liquid to a gas below the boiling point; heat is absorbed in this process. 2. Loss of water to the atmosphere; one concerning plumbers is the loss of drainage trap seal by the evaporation of the trap water.

Evaporation Rate. Rate at which a solvent evaporates.

Evaporation Rate, Final. Time interval for complete evaporation of all solvents.

Evaporation Rate, Initial. Time interval during which low boiling solvent evaporates completely.

Evaporative Condenser. Device which uses open spray or spill water to cool a condenser; evaporation of some of the water cools the condenser water and reduces water consumption.

Evaporative Cooling. A process in which outside air is pre-cooled before passing through a space; this is done by first passing the air through a layer of wet material, from which water is evaporated, increasing the water content of the air (latent heat), and reducing its dry-bulb temperature (sensible heat).

Evaporator. Part of a refrigerating mechanism in which the refrigerant vaporizes and absorbs heat.

Evaporator, Dry Type. Evaporator in which the refrigerant is in the liquid droplet form.

Evaporator Fan. Fan which increases airflow over the heat exchange surface of evaporators.

Evaporator, Flooded. Evaporator containing liquid refrigerant at all times.

Evaporator Pressure Regulator. Automatic pressure regulating valve mounted in suction line between evaporator outlet and compressor inlet; its purpose is to maintain a predetermined pressure and temperature in the evaporator.

Eviction. The legal process of ejecting a defaulting tenant from real property; unlawful detainer action.

Evidence. Testimony, documents, and objects introduced in a judicial proceeding to support the contentions of the parties.

EVT. Equi-Viscous Temperature.

EWC. Electric Water Cooler.

EWWF. Electrically Welded Wire Fabric.

EWWM. Electrically Welded Wire Mesh.

Excavation. A cavity formed by cutting, digging or scooping.

Excavation, Unclassified. Excavation paid for at a fixed price per yard, regardless of whether it is earth or rock.

Excelsior. Wood wool.

Exclusive Agency. A listing agreement with a real estate broker to sell or lease one's property during a specified period, excluding all other brokers, but reserving the owner's right to sell without paying a commission.

Exclusive Listing. An Exclusive Agency listing.

Exclusive Right to Sell. A listing agreement with a real estate broker to sell or lease one's property during a specified period, excluding all other brokers, and with the provision that the broker will receive a commission even if sold by the owner.

Excrete. To separate and expel waste matter in a continuous process.

Exculpatory Clause. A provision in a contract that relieves a party of liability.

Execute. 1. To carry out some direction, process, or work. 2. To validate a legal document by signing it.

Executive Ability. A behavioral aspect of management; the ability to manage, delegate responsibility, and coordinate all aspects of a business.

Executor. One appointed in a will to carry out its provisions.

Executrix. A female executor.

Exemplary Damages. Damages awarded to a private litigant and against a defendant to punish the wrongdoing of the defendant.

Exercise Bicycle. A stationary bicycle used for health and fitness purposes.

Exfiltration. Slow flow of air from the building to the outdoors.

Exfoliate. Removal from the surface in thin layers, flakes, or scales

Exhaust Fan. An electrical powered device to withdraw fumes, dusts, or odors from an enclosure.

Exhaust Hood. Usually a square or rectangular hood housing an exhaust fan to withdraw fumes, dust, or odors from an enclosure.

Exhaust Port. That opening which carries the fluid to the downstream pressure of a fluid system.

Exhaust Valve. A movable port which provides an outlet for the cylinder gases in a compressor or engine.

Exhaust Ventilation. Mechanical removal of air from a portion of a building.

Exhibit. A document or thing that is offered as evidence in a proceeding.

Exit. A passage or door by which one may leave a room or building; egress; a continuous and unobstructed means of egress to a public way, including intervening doorways, corridors, ramps, stairways, smokeproof enclosures, horizontal exits, exit court, and yards.

Exit Device. Panic hardware.

Exit Light. A light assembly over an exit door that is independently powered to remain lit in the event of a power failure to guide persons to safety.

Exit Lock. A dead bolt assembly mounted in an exit door.

Exit Sign. A sign located to identify the way out of a room or building.

Exothermic Reaction. A process that gives off heat.

Expanded Metal Lath. Open mesh cut and drawn from solid sheet of ferrous or non-ferrous metal; made in various patterns and metal thicknesses with uneven or flattened surface; used as a metal reinforcing for plaster.

Expanding Anchor. A soft metal sleeve, commonly lead, into which a screw-type bolt is placed to provide a stable fastener.

Expansion. An increase in size when something gets hotter.

Expansion Coefficient. The increase in length or volume per unit for a rise in temperature of 1 degree F.

Expansion Joint. 1. A vertical joint or space to allow for movement due to volume changes; also known as a relief joint or control joint. 2. A separation between two sections of concrete which is provided to allow for free movement due to temperature changes; complete separation between parts of a concrete structure; used in locations where expansion and contraction forces are anticipated; the sections are usually divided by a strip of metal, cork or bituminous material. 3. A joint through tile, mortar, concrete, or masonry down to the substrate, intended to allow for gross movement due to thermal stress or material shrinkage. 4. A device usually formed from sheet metal and having a W shaped cross section; used to provide controlled discontinuity at locations in a plaster membrane where high stresses may be encountered. 5. A structural separation between two building elements designed to minimize the effect of the stresses and movements of a building's components and to prevent these stresses from splitting or ridging the roof membrane. 6. A joint which permits pipe to expand or contract without breaking and to allow movement of the pipe caused by the pipe's expansion and contraction.

Expansion Screed. A sheetmetal plaster screed that allows expansion and contraction of adjoining panels of exterior plaster.

Expansion Shield. Device inserted in predrilled holes, usually in concrete or masonry, which expands as a screw or bolt is tightened within it, used to fasten items to concrete or masonry.

Expansion Valve. Device in a refrigerating system which reduces the pressure from the high side to the low side and is operated by pressure.

Ex Parte Proceeding. One party to a dispute appears before a judge when the other is not present.

Expendable Refrigerant System. See Chemical Refrigeration.

Experimental Analysis. An analysis procedure based on the measurement of deformations and/or strains of the structure or its model; experimental analysis may be based on either elastic or inelastic behavior.

Expert Witness. A witness who, by virtue of experience, training, skill, or knowledge of a particular field or subject, is recognized as qualified to render an informed opinion on matters relating to that field or subject.

Explosion-Proof Apparatus. Apparatus enclosed in a case which is capable of withstanding an explosion of a specified gas or vapor which may occur within it and of preventing the ignition of a specified gas or vapor surrounding the enclosure by sparks, flashes, or explosion of the gas or vapor within, and which operates at such an external temperature that a surrounding flammable atmosphere will not be ignited thereby.

Explosive. An explosive substance used to dislodge or loosen certain formations of earth and rock.

Explosive Limits. A range of the ratio of solvent vapor to air in which the mixture will explode if ignited; below the lower or above the higher explosive limit, the mixture is too lean or too rich to explode; the critical ratio runs from about one to twelve percent of solvent vapor by volume at atmospheric pressure.

Exposed Aggregate Finish. A concrete surface in which the coarse aggregate is revealed; a decorative finish for concrete achieved by removing, generally before the concrete has fully hardened, the outer skin of mortar and fine aggregates and exposing the coarse aggregate.

Exposed Grid. A framework for a suspended acoustical ceiling that is visible from below after the ceiling is completed.

Exposure. 1. The part of a felt that is not overlapped by an adjacent felt in a built-up roofing membrane; the exposure then would be that part of the felt that would be covered by the flood coat. 2. The part of a wood shingle or shake that is exposed to the weather.

Expressway. Multiple lane roadway for fast moving traffic, with access and exit ramps limited to intervals, no level crossing roads, and no pedestrian traffic allowed; also known as a Freeway.

Extender. Pigment which provides very little hiding power but is useful in stabilizing suspension, improving flow, lowering gloss, and providing other desirable qualities; generally low in cost.

Extension Ladder. A flat ladder that can be extended by sliding one section on the other to gain additional height.

Extension Link. A device used to extend the backset in a bored lock.

Exterior Door. A door which is manufactured to withstand the elements and vandalism, so it can be installed on the exterior of a building.

Exterior Elevation. An architectural drawing showing the projection on a vertical plane of an exterior surface of a building.

Exterior Fixture. An electrical lighting fixture that can be installed outdoors because of its ability to withstand the elements.

Exterior Painting. The act or process of applying paint or sealer to an exterior surface.

Exterior Plywood. A general term for plywood bonded with a type of adhesive that is highly resistant to weather, water (cold, hot, and boiling), micro-organisms, steam, and dry heat.

Exterior Tile. Tile installed in exterior applications.

Exterior Veneer. A veneer applied to weather-exposed surfaces.

Exterior Wall. Any outer wall serving as a vertical enclosure of a building other than a party wall.

Exterior Wood Door Trim. Finish components of wood such as moldings applied around openings of exterior doors.

Exterior Wood Trim. Finish components of wood such as moldings applied around openings and intersections at exterior locations.

Exterior Wood Window Trim. Finish components of wood such as moldings applied around openings of windows.

External Drive. Term used to indicate a compressor driven directly from the shaft or by a belt using an external motor; compressor and motor are serviceable separately.

External Equalizer. Tube connected to low-pressure side of a thermostatic expansion valve diaphragm and to exit end of evaporator.

External Vibrator. Vibrating device attached to formwork to consolidate fresh concrete; used primarily in precast construction.

Extinguisher, Fire. See Fire Extinguisher.

Extract. 1. To remove with effort. 2. To separate from a mixture by a physical or chemical process. 3. To separate a metal from its ore. 4. To calculate the mathematical root of a number. 5. A concentrated product.

Extraction. Obtaining a useful substance from a raw material; obtaining of a natural resource from the earth.

Extractive. Substances in wood, not an integral part of the cellular structure, that can be removed by solution in hot or cold water, ether, benzene, or other solvents that do not react chemically with wood components.

Extractive Bleeding. Stains on the surface of wood caused by extractives being leached out by water or moisture.

Extractor. A laundry appliance that removes the free water from washed fabrics by high speed centrifugal spinning, leaving them damp and ready for hot air tumble drying.

Extrados. The exterior curve in an arch or vault.

Extraordinary Flood. A deluge of greater volume than is expected in a certain place by prudent persons based on historic experience.

Extrapolate. Calculate approximate information beyond known values by projecting the trend.

Extra Work. Work performed by a contractor that is not included within the scope of the work defined by the contract documents.

Extruded Tile. A tile or trim unit that is formed when plastic clay mixtures are forced through a pug mill die of suitable configuration, resulting in a continuous ribbon of formed clay; a wire cutter or similar cut-off device is then used to cut the ribbon into appropriate lengths and widths of tile.

Extrusion. The process of squeezing a material through a shaped orifice to produce a linear element with the desired cross section; an element produced by this process.

Eye. 1. Oculus. 2. An area in the center of a tropical cyclone marked by only light winds or complete calm with no rain. 3. A metal ring through which a rope or rod is passed. 4. A loop to receive a hook.

Eyebolt, Forged. A heavy metal bolt which contains an eye at its end which may be used as a fastening device.

Eyebrow. A small roofed projection from a building.

F

FA. Fresh Air.

FAA. 1. Federal Arbitration Act. 2. Federal Aviation Administration.

Fabric, Form. Welded-wire fabric used to reinforce concrete while it is setting and gaining sufficient strength to be self-supporting.

Fabric. 1. The basic structure of a building. 2. Texture and quality, of textiles. 3. A woven cloth of organic or inorganic filaments, threads, or yarns. 4. A material that resembles cloth, like steel reinforcing mesh.

Fabrication. The construction, processing, or assembly of parts.

Fabrication, Metal. The building, construction, or manufacture of metal structures or devices.

Facade. An exterior face of a building.

Face. 1. The front or façade. 2. The exposed surface of a wall. 3. The long narrow side of a brick; the exposed surface of a wall or masonry unit. 4. The surface designed to be left exposed to view. 5. The more or less vertical surface of rock exposed by blasting or excavating, or the cutting end of a drill hole; an edge of rock used as a starting point in figuring drilling and blasting. 6. See Steel Square.

Face Amount. The sum of money stated on a check, bond, note or other instrument, exclusive of interest accumulations.

Face Board. An exterior trim board of a roof soffit system.

Face Brick. A brick selected on the basis of appearance and durability for use in the exposed surface of a wall.

Faced Wall. A wall in which the masonry facing and the backing are of different materials and are so bonded as to act together under load.

Face Frame. In cabinetwork, the front framework from which the doors are hung and drawers are inserted.

Face Layer. The outer layer of gypsum board in multilayer applications.

Face Mounted Tile. Tile assembled into units or sheets and bonded together to facilitate handling.

Face-Mounted Tile. Tile with paper applied to the face of the tile and removed with water prior to grouting.

Face-Nailing. To nail perpendicular to the surface or to the junction of the pieces joined; also called Direct Nailing.

Face of Weld. The exposed surface of a weld made by an arc or gas welding process on the side from which the welding was done.

Face Reinforcement. Reinforcement of weld at the side of the joint from which welding was done.

Face Seam. A carpet seam, either sewed or cemented, that is made without turning the entire carpet over or face-down; made during installation where it is not possible to make a back seam.

Face Shell. The side wall of a hollow concrete masonry or clay masonry unit.

Facet. A small flat surface.

Facework. Pipe valves and manifold fittings connecting the filter to the circulation lines.

Facia. See Fascia.

Facial Defect. That portion of the facial surface of the tile which is readily observed to be nonconforming and which will detract from the aesthetic appearance or serviceability of the installed tile.

Facilities Planning. Planning for the long-term use of a building, which may include furnishings, equipment, operations, maintenance, renovation, expansion, and life- cycle planning.

Facility. A property or building that is built, installed, or established to perform a particular function, such as a factory, university, or airport.

Facility Design. The architectural and engineering design of a facility.

Facility Management. The management, maintenance, and operation of a facility.

Facility Requirements. The operational needs of a facility.

Facility Services. The operational requirements of a facility including capital, personnel, materials, transportation, and public utilities.

Facing. Any masonry forming an integral part of a wall used as a finished surface; compare with Veneer.

Facing Brick. Brick made especially for exterior use with special consideration of color, texture and size.

Factor of Safety. The ratio of ultimate strength to the working stress of a material.

Factored Load. The load imposed on a member multiplied by appropriate factors, used to design reinforced concrete members.

Factory. A building or buildings containing plant or equipment for manufacturing of machinery or goods.

Factory and Shop Lumber. Lumber intended to be cut up for use in further manufacture.

Factory Edge. See Edge, 5.

Factory Finished Flooring. Flooring that has been manufactured in a factory and is ready to be installed without any further work needed on its surface.

Factory Floor. See Industrial Wood Floor.

Factory Mutual. An agency of the insurance industry that sets standards for fire safety in buildings, of which compliance is a prerequisite for fire insurance.

Factory Square. 108 square feet (10 square meters) of roofing material.

Fadeometer. Device for measuring color retention or fade resistance.

Fade-O-Meter. Mechanism used to artificially reproduce effect of sunlight on paint.

Fading. Reduction in brightness of color.

FAF. Forced air furnace

Fahrenheit. A scale for registering temperature where freezing is 32° above zero and boiling is 212°.

FAIA. Fellow of the American Institute of Architects.

Faience Mosaic. Faience tile that are less than 6 in. by 2 in. facial area, commonly 5/16 to 3/8 in. (8 to 9.5 mm) thick, and usually pre-mounted to facilitate installation.

Faience Tile. Glazed or unglazed tile, generally made by the plastic process, showing characteristic variations in the face, edges, and glaze that give a hand-crafted, non-mechanical, decorative effect.

Fail-Safe Control. Device which opens a circuit when sensing element loses its pressure.

Failure. 1. Lack of success. 2. Bankruptcy of an individual or business. 3. Unsuccessful use of a material or process. 4. Deterioration, decay, or breaking down of a material at the end of its useful life. 5. The collapse of a building or structure.

Fall. 1. The slope of land. 2. The amount of slope given to horizontal runs of pipe to provide gravity flow in the line.

Fallback. A reduction of the bitumen softening point; sometimes caused by mixing asphalt with coal-tar pitch or overheating the bitumen.

Fallout. Spray gun overspray.

False Body. Thixotropic; characteristic of paint which becomes viscous on standing but thins down on stirring.

False Set. The rapid development of rigidity in a mixed portland cement paste, mortar or concrete without the evolution of much heat; this rigidity can be dispelled and plasticity regained by further mixing without addition of water; also called Premature Stiffening and Rubber Set.

Falsework. Shoring and formwork for concrete that is removed after the concrete beams, slabs, and columns are cured.

Fan. 1. Radial or axial flow device used for moving or producing flow of gases. 2. Spacing tile joints to widen certain areas so they will conform to a section that is not parallel; also called Fanning.

Fan Coil Unit (FCU). A packaged unit consisting of a heating/cooling coil, fan, and filter, without ductwork, used to serve a space or a group of small spaces.

Fan, Exhaust. See Exhaust Fan.

Fan Light. A decorative curve topped window above the transom over an entrance door or window.

Fanning. 1. In using a spray gun, the technique of arcing; moving the spray gun away from the work. 2. See Fan, 2.

Fan Pattern. In using spray gun, the geometry or shape of spray pattern.

Fan Truss. A standard peaked roof truss configuration.

Farad. The SI unit of capacitance, such that one coulomb of charge causes a potential difference of one volt.

Faraday Experiment. Silver chloride absorbs ammonia when cool and releases it when heated; this is the basis on which some absorption refrigerators operate.

Far Face. Face of a wall farthest from the viewer; may be the outside or inside face depending on whether one is inside looking out or outside looking in.

Fascia. A flat, vertical face member or band at the surface of a building or the edge beam of a bridge, or exposed eaves of a building; also spelled Facia; a flat member of a cornice or other finish; generally the board of the cornice to which the gutter is fastened.

Fascia, Plaster. See Plaster Fascia.

Fastener. Generic term for welds, bolts, rivets, screws, and other connecting devices.

Fastener, Insulation. See Insulation Fastener.

Fastener, Pneumatically Driven. See Pneumatically Driven Fastener.

Fastener, Powder-Actuated. See Powder-Actuated Fastener.

Fast Pin Hinge. One in which the pin is fastened permanently in place; non-rising pin. See Quick Set, 2.

Fast Track. Compression of a construction schedule by over lapping some activities that otherwise would be performed sequentially.

Fat. 1. Material accumulated on the trowel during the finishing operation; often used to fill in small imperfections. 2. Describes working characteristics of a mortar containing a high proportion of cementitious material. 3. Describes working characteristics of highly plastic mortars.

Fathom. A unit of length for measuring the depth of water, equal to 6 feet.

Fatigue. A structural failure which occurs as the result of a load being applied and removed, or reversed, repeatedly over a long period of time, or a large number of cycles.

Fatigue Resistance. The ability of asphalt pavement to withstand repeated flexing or slight bending caused by the passage of wheel loads; generally, the higher the asphalt content, the greater the fatigue resistance.

Fat Mortar. Mortar containing a high percentage of cementitious components; mortar which usually does not have a sufficient amount of sand; a sticky mortar which adheres to a trowel.

Fat Mud. Mortar containing lime.

Fat Paint. Paint with too much oil.

Fatty Acid. Acid which is present in oils or fats in combination with glycerine.

Faucet. A valve for drawing liquid.

Fault. A break in the continuity of a rock formation which is caused by the shifting of the earth's crust.

Faulting. In geology, the movement which produces relative displacement of adjacent rock masses along a fracture.

Fault, Lateral Slip. See Lateral Slip Fault.

Fault, Normal. See Normal Fault.

Fault, Oblique Slip. See Oblique Slip Fault.

Fault, Reverse Thrust. See Reverse Thrust.

Fault, Strike Slip. See Strike Slip Fault.

Fault, Thrust. See Thrust Fault.

Fault Zone. Instead of being a single clear fracture, a zone may be hundreds or thousands of feet wide, consisting of numerous interlacing small faults.

Faux Marble. Fake marble; marble simulated by painting marble graining on a painted background on wood, plaster, or metal.

Feasibility Study. An analysis performed to determine the financial, economic, technical, or other advisability of a proposed project.

Feather. See Spline.

Feathered Edge. 1. The skived edge of joint tape.

Feather Edge. 1. A wood or metal tool having a beveled edge used to straighten re-entrant angles in finish plaster coat. 2. The edge of a concrete, plaster, or mortar placement such as a patch or topping that is beveled at an acute angle. 3. A tapered edge.

Featheredging Tile. A method of mitering tile by chipping away the body from beneath a facial edge of a tile in order to form a miter.

Feathering. A term used to describe the sanding or rubbing down of a surface to a feathery edge. where coating material gradually becomes thinner around the edge until it finally disappears.

Feature Strip. A narrow strip of decorated liner tile that has a contrasting color, texture, or design.

Featured Edge. A configuration of the paper-bound edge of gypsum boards which provides special design or performance features.

Featured Edge. Of gypsum board, an edge configuration of the paper bound edge that provides special design or performance.

Feeder. A conductor of a wiring system between the service equipment, or the generator switchboard of an isolated plant, and the branch circuit overcurrent device.

Feeder Duct. An enclosure for a group of electrical conductors that runs power from a central, large source to one or more secondary distribution centers.

Feel. The painter's term for the working qualities of a paint.

Feeler Gauge. A set of thin flat strips of metal of known thickness that may be used to measure small distances between surfaces.

Fee, Professional. See Professional Fee.

Fee Simple. An absolute and total interest in real property.

Feldspar. A mineral aggregate consisting chiefly of microcline, albite or anorthite.

Felt. Roofing ply sheets, consisting of a mat of organic or inorganic fibers, unsaturated, saturated, or coated with coal tar pitch or asphalt.

Felting. The process of pressing or matting together various types of hair or fibers to form a continuous fabric, known as felt.

Felt Mill Ream. The mass in pounds of 480 square feet of dry, unsaturated felt; also called Point Weight.

Female Thread. Inside threads in a pipe or fitting.

Fence. 1. A barrier used to prevent escape or intrusion or to mark a boundary, usually made of posts, boards or wire. 2. In a cast plaster shop, a plaster or clay damplaced around a model before pouring material to make the mould.

Fence Auger. A rotating drill with a screw thread used to drill deep, straight, and narrow holes for the installation of fence posts.

Fence Gate. 1. An opening in a fence. 2. A hinged or sliding panel in a fence.

Fence Hole. A hole in the ground where the main vertical support is inserted in the construction of a fence or wall.

Fence Post Hole. A cylindrical hole dug in the ground, for the insertion of a fence post.

Fence Reuse. The act or process of reusing old fence boards and other fencing materials in a new fence application.

Fender Pile. Outside row of piles that protects a pier or wharf from damage by ships.

Fenestration. The arrangement, proportioning, and design of windows and doors in a building.

Feng Shui. Chinese cultural system for sitting, arranging, and shaping buildings.

Ferroconcrete. Reinforced concrete.

Ferrous. Being iron or containing iron

Ferrous Metal. Metal alloy containing iron; ferrous pipes include wrought iron, wrought steel, rolled steel and cast iron.

Ferrous Sulphate. A green pigment commonly known as copperas.

Ferrule. 1. A metal band, ring, or cap, such as found on a tool handle to prevent splitting. 2. A cast-iron pipe fitting which, when installed in the bell of a cast-iron pipe, permits a threaded cleanout to close the opening.

Fertilizing. The act or process of adding a substance, such as manure or a chemical mixture, in order to make soil more fertile.

Festoon Lighting. An aerial span of conductors installed outdoors and supplying only weatherproof lampholders attached thereto.

FG. 1. Flat Grain. 2. Foundation Grade. 3. Fuel Gas. 4. Finish Grade.

FHMS. Flat Head Machine Screw.

FHWS. Flat Head Wood Screw.

Fiber. 1. In wood, a comparatively long, narrow, tapering wood cell, closed at both ends; also called a Tracheid. 2. An additive such as glass or cellulosic fiber to improve core flexibility and gypsum board core integrity. 3. Animal hair, sisal, manila, or glass fibers of appropriate length added to plaster mortar to increase its cohesiveness.

Fiberboard. A prefabricated sheet of compressed wood or plant fibers used for building; a homogeneous panel made from wood or cane fibers; bonding agents and other materials may be added to increase strength, resistance to moisture, fire, or decay, or to impart some other property; also called Insulation Board.

Fiberboard Insulation. Insulation sheets made from wood or cane fibers.

Fiberboard Panel. Flat sheet material made from wood that has been reduced to fibers and bonded through either a wet or dry process to obtain a specific density.

Fiberboard Sheathing. Exterior sheathing manufactured from wood or cane fibers.

Fiberboard Wall Sheathing. Flat sheet material of fiberboard secured to exterior side of exterior wall studs used to create rigidity in building superstructure and serve as base to receive siding or veneer construction.

Fibered Plaster. Basecoat plaster containing animal, vegetable, or glass fiber.

Fiberglass. 1. Glass in fibrous form used in making textiles and thermal and acoustical insulation and used for reinforcing plastics. 2. The name for products made of or with glass fibers ranging from 5 to 600 hundred-thousandths inch in diameter.

Fiberglass Insulation. A type of insulation made from glass fibers having the consistency of wool.

Fiberglass Reinforced Pipe. A pipe for liquid or gas, fabricated from glass fibers and resins for strength and durability.

Fiberglass Tank. A large container for the holding of liquids, constructed of fiberglass-reinforced plastic.

Fiber-Saturation Point. The point in the drying or wetting of wood at which the cell walls are saturated but the cell cavities are free from water; it is usually taken as approximately 23 to 30% moisture content, based on weight when oven-dry.

Fiber Saturation Point. The stage in the drying or wetting of wood at which the cell walls are saturated and the cell cavities free from water; it applies to an individual cell or group of cells, not to whole boards; it is usually taken as approximately 30 percent moisture content, based on ovendry weight.

Fibrous Concrete. A light concrete made from a fibrous aggregate, like sawdust or asbestos, for increased tensile strength and making it easy to nail.

Fiddleback-Grained Woods. Wood grain figure produced by a type of fine wavy grain found, for example, in species of maple, such wood being traditionally used for the backs of violins.

Fiduciary. A person in a relationship or position of trust; a trusted overseer.

Fiduciary Relationship. A relationship of trust and confidence between principal and agent.

Field. 1. The construction site, as contrasted to offices, factories, and work shops. 2. The surface area of a single piece of gypsum board. 3. In brick masonry, the expanse or area of wall between openings and corners, composed for the most part of stretcher units.

Field Concrete. Concrete delivered or mixed, placed, and cured on the job site.

Field Engineer. An engineer who works primarily at the jobsite as opposed to the home office; commonly represents the owner or agency and often empowered to make small engineering changes at the site to facilitate construction.

Field House. A building for athletic activities.

Field Painting. Painting at the job site.

Field Pole. Part of stator of motor which concentrates magnetic field of field winding.

Field Primed. The first coat of paint applied on the construction site rather than in the fabrication shop.

Field Tile. An area of tile covering a wall or floor; the field is bordered by tile trim.

Field Welded Truss. A truss fabricated and welded at a jobsite.

Fiery Finish. Wood finish in which hard portion of grain stands out in relief; produced by using blowtorch and stiff bristled brush.

Fifty-Fifty. A dry or dampened mixture of one part portland cement and one part extra-fine sand; this mix is used as a filler in the joints of mounted ceramic mosaic tiles to keep them evenly spaced during installation.

Figure. The pattern produced in a wood surface by annual growth rings, rays, knots, irregular coloration, and deviations from regular grain such as interlocked and wavy grain.

Filament. A conducting wire with a high melting point in an electric bulb made incandescent by an electric current.

File. A rasp-type device which is used on either wood or metal to remove burrs and rough edges or to shape the object.

Filigree. 1. Ornamental work of gold, silver, or copper as fine wire formed into delicate tracery; fine metal openwork. 2. Anything resembling this.

Fill. 1. To raise the level of land by adding earth moved in from another place or obtained by cutting. 2. An earth or broken rock structure or embankment. 3. Soil or loose rock used to raise a grade. 4. Soil that has no value except bulk. 5. Sand, gravel or other loose earth used to raise the ground level around a structure.

Filler Block. Concrete masonry unit for use in conjunction with concrete joists for concrete floor or roof construction.

Filler Metal. The metal to be added in making a welded, brazed, or soldered joint.

Filler, Joint. See Joint Filler.

Filler, Wood. See Wood Filler.

Fillet Weld. A weld at the inside intersection of two metal surfaces that meet at right angles.

Fillet Weld. A weld of approximately triangular cross section joining two surfaces approximately at right angles to each other in a lap joint, tee joint, or corner joint.

Fill, Gypsum. Troweled on plaster material to make depressions level.

Filling. In carpet making, yarns, usually of cotton, jute, or kraftcord, running across the fabric and used with the chain yarns to bind the pile tufts to the backing yarns; see Weft.

Filling In. Laying the brick or other masonry unit in the center of a wall between the face and the back.

Film Build. In painting, dry thickness characteristics per coat.

Film Former. Substance which forms skin or membrane when dried from liquid state.

Film Integrity. In paint, degree of continuity of film.

Filmogens. In paint, film-forming materials such as linseed oil and varnish resins.

Film Thickness Gauge. In painting, device for measuring film thickness above substrates; dry or wet film thickness gauges are available.

Filter. 1. Strainer; purifier. 2. A device for straining suspended particles from pool water. 3. An adjunct to air cleaning device available to serve four purposes. (1) Commercial filter - used to remove visible particles of dust, dirt, and debris, (2) Electrostatic filter - to remove microscopic particles such as smoke and haze, (3) Activated charcoal filter - to destroy odors, (4) Ultraviolet lamps or chemicals - to kill bacteria.

Filter Aid. A powderlike substance such as diatomaceous earth or volcanic ash used to coat a septum-type filter; also refers to alum as an aid to sand filtration.

Filter Block. A hollow, vitrified clay masonry unit, sometimes salt-glazed designed for trickling filter floors in sewage disposal plants.

Filter Cycle. The time of filter operation between backwash procedures; also called Filter Run.

Filter Glass. A glass, usually colored, used in goggles, helmets, and hand shields to exclude harmful light rays.

Filter Media. Fine-grain material that entraps suspended materials as they pass through the material.

Filter Rate. The rate of water flow through a filter during the filtering cycle, expressed in gallons per minute per square foot of effective filter area.

Filter Rock. Graded, rounded rock, or gravel not subject to degradation by common pool chemical used to support filter media.

Filter Run. See Filter Cycle.

Filter, Sand. See Sand Filter.

Filter Sand. A type of pool filter media composed of hard sharp silica, quartz, or similar particles with proper grading for size and uniformity.

Filter Septum. Part of a filter on which diatomaceous earth or similar filter media is deposited.

Filter Waste Discharge Piping. Piping that conducts waste water from a filter to a drainage system; connection to drainage system is made through an air gap or other approved methods.

Filtration. A method of separating a mixture made of solid particles in a liquid; after passing through a filter, the filtrate is separated from the residue.

Final Set. See Set, 4.

Final Setting Time. The time required for a freshly mixed cement paste, mortar, or concrete to achieve final set.

Financial Analysis. Three types are common. (1) comparison of current financial data with prior years, (2) comparing current financial data with that of other business in the same industry, (3) ratio analysis.

Financial Leverage. The use of borrowed money to increase the return on the investment of the owners.

Financial Statement. A balance sheet, statement of income and expense, or any other presentation of financial data.

Fine Aggregate. Concrete aggregate smaller than 1/4-inch diameter; sand.

Fine-Graded Aggregate. One having a continuous grading in sizes of particles from coarse through fine with a predominance of fine sizes.

Fine-Grained Wood. See Close-Grained Wood.

Fine Mineral Surfacing. A water-insoluble, inorganic material, more than 50% of which passes through the No. 35 sieve, that may be used on the surface of roofing material.

Fineness Modulus. An abstract number used to compare different particles or gradations of aggregate; fineness modulus is computed by adding the cumulative percentages retained on the six standard screens (#4, #8, #16, #30, #50 and #100) and dividing the sum by 100.

Fineness Modulus. An empirical factor obtained by adding the total percentages of a sample of the aggregate retained on each of a specified series of sieves and dividing the sum by 100.

Fineness of Grind. Measure of particle size or roughness of liquid paint; degree of dispersion of pigment in the binder.

Fines. Term usually pertaining to small aggregate particles capable of passing through a #200 sieve.

Fine Textured Wood. Wood with small and closely spaced pores.

Finger Joint. An end joint made up of several meshing wedges or fingers of wood bonded together with an adhesive; fingers are sloped and may be cut parallel to either the wide or narrow face of the piece.

Fingers. In airless spraying, a broken spray pattern.

Finial. An architectural ornament finishing off the top of a roof, gable, tower, canopy, or cupola.

Finish. 1. Wood products to be used in joinery, such as doors and stairs, and other fine work required to complete a building, especially the interior. 2. The process of adding stains, filler, and other materials to protect and beautify the surface of wood.

Finish Carpentry. Installation of wood finish and materials, such as doors, moldings, and window trim to a building.

Finish Coat. The final layer of any finish type material.

Finish Coat Floating. In plastering, the act of bringing the aggregate to the surface to produce a uniform texture.

Finisher. A craftsman with skill in the finishing of gypsum board joints.

Finish Grade. The final earth grade required by specifications.

Finishing Brushes. In plastering, brushes used to apply water to a smooth lime finish coat during final troweling.

Finishing Sheetrock. The taping and sanding of sheetrock seams to make ready for painting or finish.

Finishing Tools. Trowels, knives, and other special equipment required for the finishing of gypsum board joints.

Finish Paving. Finish coats of concrete, asphalt, or coated macadam on streets, sidewalks, and parking lots.

Finish Plaster. The final layer of plaster coating.

Finish Plywood. The finest grade of plywood.

Finish Shotcrete. The final coat of air-blown mortar.

Fink Truss. A three triangle symmetrical truss, commonly used in supporting large, sloping roofs.

Fir Decking. Softwood sheathing material for a deck or floor that is two inches or thicker.

Fire Alarm. An warning apparatus that can be activated to warn of fire danger.

Fire Alarm Cable. A specific electrical system cable which carries electric current to a pull station, warning horn, light, or bell in the event of a fire or other catastrophe.

Fire, Bisque. See Bisque Fire.

Fire Blocking. Intermittent solid cross-framing to retard the spread of flame within the framing cavity.

Fire Box. The interior of a fireplace or furnace, serving as combustion space.

Firebox. The portion of a fireplace that contains the fire.

Firebrick. Brick that resist high temperatures and chemical disintegration, used for lining furnaces, fireboxes, and chimneys.

Fire Brick. Brick made of refractory ceramic material which will resist high temperatures.

Fire Brick. Brick that has been tested for fire-resistance and then graded for specific construction uses; brick whose composition and characteristics make it suitable to use for masonry fireplace fire-boxes and fire chambers.

Fire Clay. An earthy or stony mineral aggregate which has as the essential constituent hydrous silicates of aluminum with or without free silica, plastic when sufficiently pulverized and wetted, rigid when subsequently dried, and suitable for use in commercial refractory products; a type of clay that can withstand great heat and is used for fire brick; also used for mortar for laying fire brick.

Firecut. A sloping end cut on a wood beam or joist where it enters a masonry wall, to allow the wood member to rotate out of the wall without prying the wall apart, if the floor or roof structure should burn through in a fire.

Fire Damper. A damper that automatically closes a duct or opening upon detection of fire.

Fire Decorating. The process of firing ceramic or metallic decorations on the surface of glazed ceramic ware.

Fire Detection Annunciator. A mechanical device which audibly announces a warning in case of a fire.

Fire Division Wall. Any wall which subdivides a building so as to resist the spread of fire.

Fire Division Wall. Any wall which subdivides a building so as to resist the spread of fire, but is not necessarily continuous through all stories to and above the roof.

Fire Door. A door which has been manufactured with specific materials and rated, designed to help hold back the spread of fire.

Fire Endurance. The ability of an assembly to perform to certain fire test criteria.

Fire Escape. Continuous unobstructed route of escape from building in case of fire, usually located on exterior of building and composed of stairs, ladders, and landings.

Fire Extinguisher. A small, portable apparatus which ejects fire-extinguishing chemicals to put out small fires.

Fire, Glost. See Glost Fire.

Fire Hydrant. A discharge fitting or apparatus with a valve and spout at which water may be drawn from the mains of waterworks; used primarily in fire-fighting, also used to service water mains and systems; also called Fire Plug.

Fire Lines. The complete wet standpipe system of buildings, including the water service, standpipe, roof manifold, Siamese connections, and pumps.

Fireplace. A masonry chamber within a building, to facilitate an open fire, with a flue to carry off smoke and fumes.

Fire Plug. Fire Hydrant.

Fireproofing. Material applied to a building element to insulate it against excessive temperatures in case of fire.

Fireproof Vault. An enclosure or room for the storing of valuables that has been constructed to withstand fire and heat.

Fire Rated Brick. Brick that has been tested for fire-resistance and then graded for specific construction uses.

Fire Rated Door. A door which has been given a rating of how long in time it can withstand fire before failure.

Fire Rated Frame. A door frame which has been given a rating on how long in time it can withstand fire before failure.

Fire Rated Panel. A panel which has been given a rating in time on how long it can withstand fire before failure.

Fire Resistance Rating. The time, in hours, or fractions of an hour, that a material or assembly will resist fire exposure.

Fire Resistance. The ability of an assembly to maintain structural stability and act as an effective barrier to the transmission of heat for a specified period of time.

Fire Resistant Sheetrock. Sheetrock which has been manufactured with fire-resistant chemicals.

Fire Resistant. Incombustible or slow to be damaged by fire; forming a barrier to the passage of fire.

Fire Retardant. A chemical treatment to reduce the ignitability of the treated material; a material or treatment which effects a reduction in flammability and in spread of fire.

Fire Retardant Chemical. A chemical preparation used to reduce flammability or to retard the spread of flame.

Fire Retardant Paint. Paint containing substance which slows down rate of combustion of flammable material or renders material incapable of supporting flame.

Fire Retardant Roof. Roofing material which has been manufactured with fire retardant chemicals.

Fire Retardant Treated Lumber. Lumber with applied or pressurized chemical treatment of lumber to retard combustion.

Fire Retardant Treated Plywood. Plywood with applied or pressurized chemical treatment to retard combustion.

Fire Separation Wall. A wall required under the building code to divide two parts of a building as a deterrent to the spread of fire; a fire wall.

Fire, Single. See Single Fire.

Fire Sprinkler Detector. A water pressure actuated switch that will signal when a sprinkler head has been activated.

Fire Sprinkler Head. The water spray device that is part of a fire sprinkler system.

Fire Sprinkler System. An arrangement of water pipes under continuous pressure that will release a water spray through spaced heads that are activated by increased temperature.

Fire Standpipe. Lines and connections within buildings that provide water source for fire fighting.

Fire Station. A building housing fire-fighting apparatus, and usually fire-fighters.

Fire Stop. 1. A projection of brickwork from walls between the joists to prevent fire from traveling between the plaster and the brick wall. 2. Material or member that seals open construction to inhibit spread of fire.

Fire Stopping. The sealing of contiguous framing cavities or areas to reduce the opportunity for rapid spread of fire and smoke; see Fire Blocking.

Fire Tapping. The taping of gypsumboard joints without subsequent finishing coats. A treatment method used in attic, plenum or mechanical areas where esthetics are not important.

Fire-Trap. A building without proper provision for escape in case of fire.

Firewall. A wall constructed to prevent or slow down the spread of fire.

Firewall, Brick. See Brick Firewall.

Fir Floor. A flooring system constructed of a softwood.

Firing. The controlled heat treatment of ceramics in a kiln or furnace, during the process of manufacture, to develop the desired properties.

Firing Range. 1. The range of firing temperature within which a ceramic composition develops properties which render it commercially useful. 2. A place where one practices the shooting of weapons.

Fir Siding. Exterior wall covering boards made from a softwood.

First Aid Kit. A packet of medical supplies for emergency use.

Fiscal Year. Financial year; an accounting period of 12 months.

Fished. A means of installing electrical wiring in existing inaccessible hollow spaces of buildings with a minimum damage to the building finish.

Fished Joint. An end butt splice strengthened by pieces nailed on the sides.

Fish Eyes. A term used to describe small blemishes occasionally found in lime finish coats; approximately 1/4 inch in diameter, they are caused by lumpy lime.

Fishmouth. An opening at the exposed edge of a roofing ply sheet where the asphalt bond is lacking or the felt is wrinkled.

Fish Oil. The only animal oil used to any extent in the paint industry; extracted from fish such as sardine, menhaden, and pilchard.

Fish Tape. A steel tape used by electricians to push through an electrical conduit and pull back the conductors being installed.

Fissure. A narrow crack, usually long and deep, caused by breaking or parting

Fitting. 1. A device used for connecting pipes together. 2. An accessory such as a locknut, bushing or other part of a wiring system which is intended primarily to perform a mechanical rather than an electrical function.

Fitting Gain. The space inside a fitting required by a pipe.

Fixative. A protective coating applied to drawings in crayons, pastel, charcoal, and pencil, usually by spraying, to prevent colors from rubbing off.

Fixed. Attached or fastened in place by nails, screws, bolts, conduit, piping systems, or other means.

Fixed Assets. Assets of a more or less permanent nature whose useful life is more than one year, such as fixtures, equipment, real estate, and trucks.

Fixed Displacement Pump. A pump in which the displacement per cycle cannot be varied.

Fixed Door. A door, of a pair, that is normally kept in the closed position.

Fixed End Beam. A beam fixed upon a support which prevents its rotation.

Fixed Fee. A stipulated lump sum fee.

Fixed Liabilities. Liabilities that are carried for over a one year period, such as mortgage, and equipment payments.

Fixed Window. Sash or glass that is immovably mounted in a wall.

Fixture Branch. A pipe connecting several fixtures with the water supply.

Fixture Carrier. A mounting frame built into a wall to support a plumbing fixture.

Fixture Carrier. A wall mounted frame for mounting and support of a plumbing fixture.

Fixture Drain. The drain pipe from the trap of a fixture to where the pipe joins with any other drain pipe.

Fixture, Lighting. See Lighting Fixture.

Fixture, Plumbing. See Plumbing Fixture.

Fixture Unit. A design factor to determine the load-producing value from a given fixture so that the drainage piping is large enough to carry the liquids and wastes; for instance, the unit flow rate from fixtures is assumed to be one cubic foot or 7.5 gallons of liquid per minute; fixtures are rated as multiples of this unit of flow.

Flag. 1. See Flagstone. 2. End of hog brush bristle which divides into two or more branches like a tree; flagging provides brush with ability to hold paint.

Flagging. 1. Flagstone paving. 2. See Flag, 2.

Flagpole. A pole used for raising and displaying flags.

Flagpole Outrigger. A projecting pole run out from a structure to provide a flagpole.

Flagstone. A hard flat, usually rectangular, stone slab used for paving; also called a flag.

Flake. A small flat wood particle of predetermined dimensions, uniform thickness, with fiber direction essentially in the plane of the flake; in overall character resembling a small piece of veneer; produced by special equipment for use in the manufacture of flakeboard.

Flakeboard. A type of particleboard composed of flakes bonded together with a synthetic resin of other suitable binder.

Flaked. Irregularities left on the edge of the tile mainly due to the use of machine cutting tools.

Flaking. Detachment of small pieces of paint film.

Flameproof Tile. Tile that has been made resistant to flame.

Flameproofing. An act or process of making an object resistant to the action of flame.

Flame Spread Classification. A standard measurement of the relative surface burning characteristics of a building material when tested.

Flame Spread Rating. A measure of the rapidity with which fire will spread across the surface of a material.

Flame Spread. The propagation of a flame away from the source of ignition across the surface of a liquid or a solid, or throughout the volume of a gaseous mixture.

Flame Test for Leaks. Tool which is principally a torch; when a halogen mixture is fed to the flame, this flame will change color in the presence of heated copper.

Flammable. Capable of being easily ignited; same as inflammable; the opposite is nonflammable.

Flammable Liquids. Liquids having a flash point below 140° F. (60° C.) and a vapor pressure not exceeding 40 psia (2.81 kg/cm^2) at 100° F. (38° C.).

Flange. A rib or rim on an object for strength, for guiding, or for attachment to another object.

Flanged Pipe. A pipe that has been fitted with a projecting ring, ridge or collar, to strengthen, prevent sliding, or accommodate attachments.

Flanking Paths. Paths by which sound travels around an element that is intended to impede it.

Flapper Valves. Thin metal valve used in refrigeration compressors which allows gaseous refrigerants to flow in only one direction.

Flare. An enlargement at the end of a piece of flexible tubing by which the tubing is connected to a fitting or another piece of tubing; this enlargement is made at about a 45 degree angle; fittings grip it firmly to make the joint leakproof and strong.

Flare Header. 1. A brick burned on one end to a darker color than the face. 2. A header of darker color than the field of the wall

Flare Nut. Fitting used to clamp tubing flare against another fitting.

Flashback. In welding. a recession of the flame into or back of the mixing chamber of the torch.

Flash Cove. A detail in which a sheet of resilient flooring is turned up at the edge and finished against the wall to create an integral baseboard.

Flash Gas. Instantaneous evaporation of some liquid refrigerant in evaporator which cools remaining liquid refrigerant to desired evaporation temperature.

Flashing. 1. A thin, continuous sheet of impervious material such as metal, plastic, rubber, or waterproof paper used to prevent the passage of water through a joint in a wall, roof, or at a chimney; the material used and the process of making watertight the roof intersections and other exposed places on the outside of a structure. 2. Nonuniform appearance on surfaces in which coating dries with spotty differences of color or gloss.

Flashing Block. In masonry, metal flashing used to block a parapet wall and prevent roof leaks around such a wall.

Flashing Cement. A plastic mixture of cutback bitumen and mineral stabilizers, asbestos fibers or other inorganic fibers, used for reinforcing, mixed with a solvent and used as an adhesive.

Flashing, Roof. See Roof Flashing.

Flashing, Step. See Step Flashing.

Flashing, Through-Wall. See Through-Wall Flashing.

Flash Point. The temperature at which the material gives off flammable vapor in sufficient quantity to ignite momentarily on the application of a flame under specified conditions.

Flash Set. The rapid development of rigidity in a mixed portland cement paste, mortar, or concrete usually with the evolution of considerable heat, which rigidity cannot be dispelled nor can the plasticity be regained by further mixing without addition of water; also referred to as Quick Set or Grab Set.

Flash Weld. Resistance type weld in which mating parts are brought together under considerable pressure while a heavy electrical current is passed through the joint to be welded.

Flat Arch. An arch or span constructed with a flat top and bottom; steel angle iron is usually used to help support this arch.

Flat Asphalt. A roofing asphalt that has a softening point of approximately 170° F. (77° C).

Flat Finish. Dull finish, no gloss.

Flat-Grained Wood. Lumber that has been sawed parallel to the pith and approximately tangent to the growth rings; lumber is considered flat-grain when the annual growth rings make an angle of less than 45 degrees with the surface of the piece.

Flat Head Screwdriver. See Slotted Screwdriver.

Flat Head Wood Screw (FHWS). A wood screw with a countersunk head flush with the surface, driven with a screwdriver.

Flat Mill. A kind of grinding mill used to grind paint pigments; the mill consists of two stones, a lower stone which revolves and an upper stone which is stationary.

Flat Paint. An interior paint with a high proportion of pigment; it dries to a flat, or lusterless, finish.

Flatplate Collector. The most common and often-used collector in a solar system; the most visible portion of a solar system appearing on the roof or in the yard; device in which sunlight is converted to heat on a plane surface usually made of metal or plastic; a heat transfer fluid circulates through the collector to transport heat to be used directly or to be stored.

Flat Position Weld. The position of welding wherein welding is performed from the upper side of the joint and the face of the weld is approximately horizontal.

Flat Seam. A sheet metal roofing seam that is formed flat against the surface of the roof.

Flat Slab. A reinforced concrete slab that is designed to span, without any beams or girders, in two directions to supporting columns; a 2-way slab supported at its corners only.

Flat Stretcher Course. A brick course of stretchers that are set on edge and expose their flat sides on the surface or face of the wall.

Flattening Agent. An ingredient, usually a metallic soap, such as calcium, aluminum or zinc stearate, used in lacquers and varnishes to reduce the gloss or to give a rubbed appearance; also called Flatting Agent.

Flattening Oil. A varnish-like composition, made of thickened oil dissolved in a thinner, used to reduce paste paint to a flat paint; also called Flatting Oil.

Flatting. Loss of gloss in coating film.

Flatting Agent. See Flattening Agent.

Flatting Oil. See Flattening Oil.

Flat Trowel. The flat trowel is used in conjunction with the hawk for the transferring of mortar from the mortarboard to the wall or to other vertical surfaces; frequently used for spreading pure cement on the finished floor coat; also used for spreading mortar on floor surfaces before tiles are set; commonly used to put the finish on a freshly poured concrete floor.

Flat Varnish. Varnish which dries with reduced gloss, made by adding such materials as silica, wax, or metallic soaps to the varnish.

Flat Wall Paint. A type of interior paint which is designed to produce a flat or lusterless finish.

Flat Washer. A washer which goes under a bolt head or a nut to spread the load, prevent loosening, and protect the surface.

Flatwork. Concrete slabs on grade.

Flecks. See Wood Rays.

Flemish Bond. A brick bond consisting of headers and stretchers alternating in every course, so laid as always to break joints, each header being placed in the middle of the stretchers in courses above and below.

Flemish Cross Bond. A brick bond that has alternate courses of Flemish headers and stretcher courses. The headers are directly over each other and the alternate stretcher courses are crossed over each other.

Flemish Garden Bond. A brick bond where bricks are laid so that each course has a header to every 3 or 4 stretchers.

Flex. Flexible electrical conduit.

Flexibility. 1. The ability to bend or conform to new conditions. 2. The ability of an asphalt pavement structure to conform to settlement of the foundation; generally, flexibility of the asphalt paving mixture is enhanced by high asphalt content.

Flexible. Limber, bendable; not stiff; the opposite of rigid.

Flexible Conduit. Electrical conduit made of a spirally wound metallic strip.

Flexible Cord. A flexible insulated electrical cable having a plug at one or both ends used to connect an appliance or other fixture to a receptacle, manufactured to resist tangling.

Flexible Coupling. A mechanical connection that adapts to misalignment between moving parts.

Flexible Ductwork. Flexible ductwork manufactured in various diameters, made from spiral wire covered in plastic and commonly insulated, for use in the transfer of air in heating, cooling and ventilating systems.

Flexible Wiring. Electrical wiring that permits movement from expansion, contraction, vibration, or rotation.

Flexural Strength. 1. A property of a material or structural member that indicates its ability to resist failure in bending. 2. The resistance of a specified sample size of a gypsum board to failure caused by a transverse load.

Flexure Formula. A formula for determining values for the design of members or elements subjected to bending.

Flint Paper. Abrasive paper which is grayish-white in color; inexpensive but has short working life.

Flitch. A portion of a log sawn on two or more faces, commonly on opposite faces leaving two waney edges; when intended for resawing into lumber, it is resawn parallel to its original wide faces; or, it may be sliced or sawn into veneer, in which case the resulting sheets of veneer laid together in the sequence of cutting are called a flitch

Flitch-Sliced Veneer. A thin sheet of wood cut by passing a block of wood vertically against a long, sharp knife.

Float. 1. A tool or apparatus for smoothing a surface, used by plasterers and concrete masons. 2. A sum of money in excess of daily needs. 3. See Slack, 1.

Float Coat. The final mortar coat over which the neat coat, pure coat, or skim coat is applied.

Float Finish. A finish coat that is rough from aggregate material found in the plaster mortar.

Float Glass. Glass sheet manufactured by cooling a layer of liquid glass on a bath of molten tin.

Floating. 1.Separation or layering of pigment in mixture of pigment. 2. A method of using a straightedge to align mortar with the float strips or screeds; this technique also is called dragging, pulling, rodding, or rodding off. 3. The rising of a swimming pool out of the ground, caused by water pressure under the pool; this only occurs when the pool is not filled with water.

Floating Angle. A method of applying gypsumboard designed to allow structural movement at interior corners.

Floating Edge. In gypsumboard installation, an actory edge applied in such a manner that the edge does not lie directly over a framing member and is unsupported.

Floating Joint. In gypsum board installation, a condition where the butt joint does not lie directly over a framing member; floating joints should be back-blocked.

Float Strip. A strip of wood about 1/4 inch thick and 1-1/4 inch wide, used as a guide to align mortar surfaces.

Float Switch. An electrical switch that is actuated by the rising or falling of a float in liquid.

Float Trap. A floating device in a plumbing fixture which opens or closes a valve to prevent sewer air gases from escaping back through the fixture.

Float Valve. Type of valve which is operated by sphere or pan which floats on liquid surface and controls level of liquid.

Floc. Gel-like substance formed when coagulant, usually alum, combines with suspended alkaline matter in pool water and precipitates out.

Flocculate. Agglomeration of undispersed pigment particles.

Flocculating Agent. Compound that coalesces finely suspended particles.

Flocculent. A compound, usually some type of alum, used with sand-type filters to form a thin layer of gelatinous substance on the top of the sand; aids in trapping fine suspended particles which might pass through the floc.

Flocked Carpet. Single-level velvety pile carpet composed of short fibers embedded on an adhesive-coated backing.

Flocked Paper. Wallpaper covered with flocking.

Flock Finish. Finish obtained by spraying or sifting flock (short fibers of wool, silk, rayon) onto a surface to which the flock fibers will adhere.

Flocking. Very short or pulverized fiber used to form a velvety pattern or design on a surface.

Flococoating. Process of finishing by flowing finishing material on article by means of hose, allowing excess to drain into tank.

Flood. An overwhelming volume of water by overflowing of a body of water; a deluge.

Flood Coat. The top layer of bitumen of a built-up roofing membrane; usually it is aggregate surfaced; if the flood coat is applied correctly it is poured, not mopped, to a weight of 60 pounds per square for asphalt and to a weight of 75 pounds per square for coal-tar pitch.

Flooded System. Type of refrigerating system in which liquid refrigerant fills most of the evaporator.

Flooded System, High-Side Float. Refrigeration system which has a float operated by the level of the high-side liquid refrigerant.

Flooded System, Low-Side Float. Refrigerating system which has a low-side float refrigerant control.

Flooding. Same as floating.

Flooding. Act of allowing a liquid to flow into a part of a system.

Flood Lamp. A strong source of artificial illumination which projects a broad beam.

Flood Level Rim. See Flood Rim.

Floodlight. Artificial lighting in a broad beam.

Floodplain. Level land that may be submerged by floodwaters.

Flood Rim. The highest point that water can reach within a plumbing fixture without overflowing; also called Flood Level Rim.

Floor. The lower interior surface of a room in a building.

Floor, Access. See Access Floor.

Floorboard. One of various types of boards used for flooring.

Floor Box. A metal electrical rough-in box fed by conduits in or under the floor to provide for a floor outlet.

Floor Brick. Smooth dense brick, highly resistant to abrasion, used as finished floor surfaces.

Floor Closer. A closing device installed in the floor under a door.

Floor Deck. Sheet steel formed to fluted or ribbed profile to span between supports (usually joists or beams) to support floor system (usually concrete slab) and live loads.

Floor Drain. 1. A drain in a floor which connects to the plumbing system and removes unwanted water from an area or room. 2. An opening or receptacle located in a concrete floor, connected to a trap to receive the discharge from indirect waste.

Floor Finish. The stain, paint, wax, or polish on a floor.

Floor Finishing. The act or process of the final sanding, coloring, and sealing of a wooden floor system.

Floor Grating. Open grid of metal bars structurally formed.

Floor, Gym. See Gym Floor.

Floor Hardener. A chemical applied to concrete floors to protect them from wear.

Floor Hinge. A combined pivot hinge and closing device set either in the floor or in the bottom of the door; it may be spring type only or may be combined with liquid control.

Floor, Industrial Wood. See Industrial Wood Floor.

Flooring. The material with which a floor is surfaced, such as carpeting, wood strips, wood parquet, linoleum, or sheet vinyl.

Flooring, Marble. See Marble Flooring.

Floor Joist. A support beam, commonly installed in parallel with other beams to create a structural floor system, after which floor sheathing is fastened.

Floor Level. The elevation of any floor in a building.

Floor, Level. See Level Floor.

Floor Masonry. Shaped or molded masonry units such as, stone, ceramic brick, tile or concrete used for finished flooring.

Floor Mat Frame. A frame, usually metal, to contain a floor mat and provide a means to anchor the mat to the floor.

Floor Mat. A piece of coarse woven or plaited fabric used as a floor covering.

Floor Molding. A piece of trim at the juncture of the wall and floor.

Floor Mounted Bench. A work surface that has its support legs mounted to the floor.

Floor Mounted Transformer. A transformer that has its supports mounted to the floor to prevent vibration and movement.

Floor Mud. Mortar which does not contain lime, used to float horizontal surfaces; also called Deck Mud.

Floor Patch. Material used to repair damage to a floor.

Floor Pedestal. A member, such as a short pier, used as a base for a floor system.

Floor Plate. Steel or aluminum plate with a non-skid design formed on one side for use in constructing industrial floors, platforms, landings, and stairs.

Floor Price. The least amount that an informed seller would accept for leasing or selling property.

Floor Removal. The act or process of the tearing up and carrying away an old floor system.

Floor Safe. A place or receptacle to keep valuables, installed in a floor.

Floor Sink. A small sink, typically 12" x 12", set in the floor to receive drainage such as condensation water.

Floor Slab. A reinforced concrete slab on grade or elevated, used as a floor.

Floor, Terrazzo. See Terrazzo Floor.

Floor, Tile. See Tile Floor.

Floor Tile, Clay. See Clay Floor Tile.

Floor Varnish. A varnish made specifically for application to floors.

Floor Wax. A substance spread onto flooring that seals, protects, and can be polished to a shine.

Flow. A measure of self leveling.

Flow After Suction. Flow of mortar measured after subjecting it to a vacuum produced by a head of two inches of mercury;

the suction apparatus and its use is described in Sections 27 and 28 of ASTM C 91.

Flow Gradient. A drainageway slope determined by the elevation and distance of the inlet and outlet and by required volume and velocity.

Flowing Varnish. A varnish which has been designed to produce a smooth lustrous surface without rubbing or polishing.

Flow Meter. Instrument used to measure velocity or volume of fluid movement.

Flow of Mortar. Measure of mortar consistency, sometimes termed the Initial Flow, determined on the flow table described in ASTM C 230; use of the flow table and method of calculating the flow is described in Section 9 of ASTM C 109.

Flue. A passage for smoke and combustion products from a furnace, stove, water heater or fireplace; the opening in a chimney through which smoke passes.

Flue Liner. Heat-resistant firebrick or other fire clay materials that make up the lining of a chimney.

Fluffing. Lint and fuzz that appears on newly installed carpet, which is merely the factory-sheared pile ends working their way to the surface, not the tufts or pile yarns themselves; this condition disappears as the carpet is used; also called Shedding.

Fluid. Substance in either a liquid or gaseous state; substance containing particles which move and change position without separation of the mass.

Fluid Adjusting Screw. A screw on a spray gun which controls the amount of fluid entering the gun.

Fluid Applied Elastomer. An elastomeric material, which is fluid at ambient temperature, that dries or cures after application to form a continuous membrane.

Fluid Applied Roof Membrane. A roof membrane applied in one or more coats of a liquid that cure to form an impervious sheet.

Fluid Applied Roof. A roof coated with an asphalt-based liquid.

Fluid Flow. A measure of flow through a spray gun with atomizing air shut off.

Fluid Hose. Specially designed hose for paint materials; usually black.

Fluid Nozzle. Spray gun fluid tip with orifice; the needle and tip combination.

Fluorescent Fixture. An illuminating device which uses a tubular electric lamp which has a coating of fluorescent material on its inner surface and contains mercury vapor.

Fluorescent Lamp Starter. See Ballast.

Fluorescent Paint. Luminous paint which glows only during activation by ultraviolet or black light.

Fluorite. Fluorspar, CaF_2, an inorganic mineral of the isometric form, used as a source of fluorine for fluxing of glasses and glazes.

Flush. 1. Adjacent surfaces even, or in the same plane. 2. Operation to remove any material or fluids from refrigeration system parts by purging them to the atmosphere using refrigerant or other fluids. 3. To operate a water closet or urinal valve.

Flush Bolt. A door bolt so designed that when applied it is flush with the face or edge of the door.

Flush Bushing, Copper. A threaded pipe fitting used to connect two pipes with different diameters together and create a continuous smooth surface, made of copper.

Flush Cabinet Construction. A method of building cabinets where the fronts of the doors and drawers are flush with the face frame.

Flush Door. A door with a smooth surface and no protrusions.

Flushing. Of asphalt paving, see Bleeding, 1.

Flush Molding. A piece of trim which is fastened flat onto or flush with a wall and sometimes is used as a divider or chair rail.

Flushometer. A high pressure flush valve for a toilet or urinal.

Flush Tank. The water tank or cistern on a toilet to supply flushing water.

Flush Valve. A regulated water valve at a toilet or urinal to supply flushing water directly from a high pressure water line.

Fluted Column. Column with vertical decorative, semi-circular channeled shafts.

Flux. 1. A substance that promotes fusion in a given ceramic mixture. 2. A chemical used when soldering, brazing, or welding to prevent oxides from forming when heated. 3. Fusible material used in welding or oxygen cutting to dissolve and facilitate removal of oxides and other undesirable substances. 4. The density of a magnetic field.

Fly Ash. Extremely fine ash from the burning of pulverized coal; when it has over 90 percent silica content it may be used as a pozzolan in concrete; it can improve the concrete by increasing the volume of cement paste.

Fly Gallery. The space above a theater stage where the scenery is suspended.

Flying Buttress. A masonry buttress springing from a separate column, forming an arch with the wall it supports

Flying Formwork. Large sections of slab formwork that are moved by crane.

FM. Factory Mutual.

Foam Concrete. See Cellular Concrete.

Foamed Concrete. Concrete made very light and cellular by the addition of a prepared foam or by generation of gas within the unhardened mixture.

Foam Extinguisher. A extinguisher which releases chemical foam to put out fires.

Foam Gasket. Joint sealing material made of rubber or plastic foam strips.

Foamglass Insulation. A thermal insulation made by foaming glass with hydrogen sulfide; it is manufactured in the form of block or board and has a low fire hazard rating.

Foaming. 1. Frothing. 2. Formation of a foam in an oil-refrigerant mixture due to rapid evaporation of refrigerant dissolved in the oil; this is most likely to occur when the compressor starts and the pressure is suddenly reduced.

Foam Leak Detector. System of soap bubbles or special foaming liquids brushed over joints and connections to locate leaks.

Foam Plaster Base. A rigid type foamed backing which acts as a plaster base.

FOB. Free On Board; without charge for delivery to and placing on board a carrier at a specified point; FOB Jobsite means material cost includes cost of delivery to jobsite.

Focal Depth. Depth of the earthquake focus (or hypocenter) below the ground surface.

Focus. The point of origin of an earthquake; the point at which the rupture occurs; synonymous with the hypocenter.

Fog Curing. 1. Storage of concrete in a moist room in which the desired high humidity is achieved by the atomization of fresh water. 2. Application of atomized fresh water to concrete, stucco, mortar, or plaster.

Fogging. Misting.

Foil. 1. Metal hammered into a thin sheet. 2. A cusped element in Gothic tracery.

Foil Back. See Foil Backed Gypsum Board.

Foil Backed Gypsumboard. Gypsumboard with aluminum foil laminated to its back surface as a vapor retardant and thermal insulator. Also called Foil Back.

Folded Plate. A roof structure in which strength and stiffness derive from a pleated or folded geometry; special class of shell structure formed by joining flat, thin slabs along their edges so as to create a three dimensional spatial structure.

Folding Door. A door with hinged leaves, often provided with a ceiling or floor track.

Folding Stair. A hinged stair, attached to and concealed within the ceiling, which can be raised and lowered.

Folio. A page number referring to the disposition of source of an entry or posting in the books of the business.

Font. A receptacle in a church for baptismal water or holy water.

Food Preparation. The action or process of making food ready for cooking and consumption.

Food Service Equipment. Appliances, fixtures, and materials used in the cooking and serving of food in large quantities.

Foot. A unit of length equal to 12 inches or 30.48 centimeters.

Foot-Candle. The illumination produced by one lumen of luminous flux spread uniformly over an area of one square foot.

Foot Grille. Metal grating usually located in floor near building entrance to allow for scraping of dirt, ice, and snow from footwear before proceeding to building interior.

Footing. A masonry section, usually concrete, in a rectangular form wider than the bottom of the foundation wall or pier it supports.

Footing Drain. 1. Drain around the perimeter of a building, at the footings, to drain groundwater or rainwater away from the building. 2. Drain at the footing of a retaining wall to prevent accumulation of hydrostatic pressure that could collapse the wall.

Footing Excavation. The temporary removal of earth for installing a footing.

Footing Form. A wooden or steel structure, placed around the footing that will hold the concrete to the desired shape and size; also called Footing Formwork.

Footing Formwork. See Footing Form.

Footing Reinforcing. The placing of metal or steel bars in a freshly poured or placed concrete footing to strengthen it.

Foot-Pound. A unit of work equal to the work done by a force of one pound acting through a distance of one foot in the direction of the force.

Foot-Poundal. An absolute unit of work that is equal to the work done by a force of 1 poundal in moving a body through a distance of 1 foot, equivalent to about 0.04 joules.

Footprint. The outline of a building on the ground, used in site planning and in judging compliance with planning and zoning laws.

Foots. Dregs; settlings in vegetable oils.

Force. Energy exerted on an element tending to cause motion; magnitude, direction and point of application are all used to describe a force.

Force Account. Construction work on a time and material basis.

Forced Air. Air blown by a fan from a furnace or air conditioner.

Forced Air Furnace. An appliance, with a heat exchanger and ventilating fan, that supplies hot air for space heating.

Forced Convection. Movement of fluid by mechanical force such as fans or pumps.

Forced Draft. In a boiler, combustion air forced by a blower in the burner section through the boiler and up the stack.

Forced Drying. Acceleration of drying by increasing the temperature above ambient temperature using an oven, infrared lamp, or other heat source.

Force-Feed Oiling. Lubrication system which uses a pump to force oil to surfaces of moving parts.

Force Majeure. An overwhelming, but unanticipated event.

Foreclosure. A legal procedure for selling a mortgaged property in default to satisfy the debt.

Forehand Welding. A gas-welding technique wherein the flame is directed toward the progress of welding.

Foreman. An experienced worker who works with and usually leads a crew or gang.

Forensic Architecture. A science and practice that deals with application of architecture and construction facts and scientific methods to construction and legal problems.

Forensic Engineering. A science and practice that deals with application of engineering and construction facts and scientific methods to construction and legal problems.

Forge. 1. A furnace or hearth for melting or refining metal. 2. A furnace where metal is heated and wrought.

Forged Eyebolt. A heavy metal bolt which contains an eye at its end which may be used as a fastening device.

Forged. Formed by heating and hammering.

Forged Steel. Wrought steel.

Fork Lift. A vehicle with a horizontal fork in front for lifting and moving loads; a fork lift truck.

Form. Temporary structure built to contain concrete while it sets; also called Formwork.

Form Deck. Thin, corrugated steel decking that serves as permanent formwork for a reinforced concrete deck.

Form Fabric. Welded-wire fabric used to reinforce concrete while it is setting and gaining sufficient strength to be self-supporting.

Form-Fit Area. See Free Form.

Formica. Brand name and trademark for any of the various laminated plastic products, usually used for surface finish on cabinets or millwork.

Formica Partition. A dividing wall finished with a layer of Formica.

Forming. 1. The act of installing wooden forms to accept reinforcing steel and concrete. 2. The shaping or molding of ceramic ware.

Form Lacquer. Thin lacquer or varnish used to coat concrete forms to prevent concrete from adhering to the forms.

Form Oil. Oil applied to the interior surface of formwork to promote easy release from the concrete when forms are removed.

Form Release. A substance applied to concrete forms to make stripping after pouring easier.

Form Tie. A steel rod with fasteners on each end, used to hold together the formwork for a concrete wall.

Form, Slab. See Slab Form.

Formwork. See Form.

Forsterite Porcelain. A vitreous ceramic whiteware for technical applications in which forsterite, Mg_2SiO_4, is the essential crystalline phase.

Fossil Fuel. A natural fuel such as coal, gas, or petroleum formed in the geological past from the remains of living organisms.

Fossil Resin. Any of the natural or earth type resins, such as kauri and the Congo copals, which derive their characteristics through aging in the ground.

Foundation. The portion of a building that has the sole purpose of transmitting structural loads from the building into the earth. That part of a building or wall which supports the superstructure.

Foundation Backfill. Earth or earthen material displaced during excavation and replaced around the constructed foundation.

Foundation Block. Masonry block used for foundation work, commonly denser than regular block.

Foundation Investigation. See Geotechnical Investigation.

Foundation Mat. A grid of reinforcement steel for concrete foundations.

Foundation Reinforcing. The placing of metal or steel bars in a freshly poured or placed concrete foundation to strengthen it.

Foundation Resin. Piping around the base of a building to collect water and convey to the drainage system.

Foundation Soil. See Subgrade.

Foundation Vent. Screened opening below the floor line to provide natural ventilation to the foundation crawl spaces.

Foundation Wall. That portion of a load-bearing wall below the level of the adjacent grade, or below the first tier of floor beams or joists, which transmits the superimposed load to the footing.

Foundry. A workshop for casting metals.

Fountain. 1. A jet or jets of water made to spout for ornamental purposes or for drinking and the structure for this purpose. 2. A natural spring of water.

Fountain, Wash. See Wash Fountain.

Four Dimensional. Involving four dimensions, consisting of elements requiring four coordinates to determine them; the space-time continuum.

Four Pole Motor. 1800 rpm, 60 Hz electric synchronous speed motor.

Fourth Dimension. Time, in addition to height, width, and depth.

Foyer. An anteroom, lobby, entrance hallway, or vestibule.

FPL. Forest Products Laboratory.

Fraction. 1. A numerical quantity that is not a whole number, as in 1/2 or 0.5; a number expressed as one number divided by another, as in 2/3 or 4/5; the upper number is the numerator and the bottom number is the denominator. 2. A portion of a mixture separated by distillation.

Fracture. 1. A break, usually resulting in actual separation of the material; in structures, the characteristic result of tension failure. 2. See Crack, 1.

Frame. 1. An enclosing border as in a picture frame. 2. The surrounding or enclosing woodwork, as around windows or doors. 3. The skeleton of a building; that is, the rough structure of a building, including interior and exterior walls, floor, roof, and ceilings. 4. To form together or construct large assemblies from smaller components. 5. A structural system consisting of relatively long, prismatic members fastened together; a rigid frame is one in which the joints can transmit moments as well as forces and which therefore does not require a braced frame for rigidity.

Frame High. To construct the brickwork up to the top of the door or window frame; the lintel is then laid across the opening and rests upon the brickwork on each side of the frame.

Frame, Balloon. See Balloon Frame.

Frame, Braced. See Braced Frame.

Framed Connection. 1. One that is capable of resisting moments. 2. A shear connection between steel members made by means of steel angles or plates connecting to the web of the beam or girder.

Frame, Door. See Door Frame.

Frame, Ductile Moment Resisting Space. See Ductile Moment Resisting Space Frame.

Frame, Moment Resisting. See Moment Resisting Frame.

Frame, Platform. See Platform Frame.

Framer. 1. A carpenter who constructs wood framing. 2. A carpentry contractor. 3. One who frames pictures and makes their frames.

Frames. Racks at the back of a Jacquard loom, each holding a different color of pile yarn; in Wilton carpets, 2 to 6 frames may be used and the number is a measure of quality as well as an indication of the number of colors in the pattern, unless some of the yarns are buried in the backing.

Framing. The rough wooden structural skeleton of a building, including interior and exterior walls, floor, roof, and ceilings.

Framing, Ceiling. See Ceiling Framing.

Framing, Door. See Door Framing.

Framing Lumber. Wood members of framing systems which are manufactured by sawing, resawing, passing lengthwise through standard planing machine, crosscutting to length, and matching, but without further manufacturing.

Framing Member. The stud, plate, joist, or furring component to which the exterior and interior surfacing materials are attached; normally made of wood or metal.

Framing, Roof. See Roof Framing.

Framing, Timber. See Timber Framing.

Framing, Wall. See Wall Framing.

Franchise Tax Board. In California, a department of state government that collects taxes from individuals and businesses.

Fraud. A false statement of fact that is designed to deceive.

Free and Clear. Real property that has no liens or encumbrances.

Freeboard. Distance between poolwater surface level and deck level.

Free Body Diagram. A diagram, or drawing, in which on element of structure is isolated from its surroundings, and the effect of its surroundings is shown only as forces; see Vector, 1.

Freeform. Freely adapted poolshape.

Free Form. A floor area, usually in a department store or salon, not bounded by walls and of nonrectangular shape; sometimes called Form-Fit Area.

Freehand Floating. The application of wall mortar without the use of guide screeds; this technique is used by specialists when they are setting glass mosaic murals.

Freestone. Fine-grained sedimentary rock that has no planes of cleavage or sedimentation along which it is likely to split.

Free Water. All water contained by gypsum board, concrete, mortar, or plaster in excess of that chemically held as water of crystallization;

Freeway. See Expressway.

Freezer Alarm. A bell or buzzer used in many freezers which sounds an alarm when freezer temperature rises above safe limit.

Freezer Burn. Condition applied to food which has not been properly wrapped and that has become hard, dry, and discolored.

Freeze-Up. 1. Formation of ice in the refrigerant control device which may stop the flow of refrigerant into the evaporator. 2. Frost formation on an evaporator which may stop the airflow through the evaporator.

Freezing. Change of state from liquid to solid.

Freezing Point. Temperature at which a liquid will solidify upon removal of heat. The freezing temperature for water is 32° F. (0° C.) at atmospheric pressure.

Freight Elevator. A hoisting device to raise or lower cargo.

Freight-In. A merchandising account that records the transportation costs on purchases paid by the buyer.

Freize Carpet. A rough, nubby-textured carpet using tightly twisted yarns; a carpet with pile of uncut loops or a combination of cut and uncut loops.

Freize Yarn. A tightly twisted yarn that gives a rough, nubby appearance to the pile; in addition to use in plain colors, it is employed to form designs against plain grounds and thus gives an engraved effect.

French Curve. A transparent plastic drafter's tool for drawing miscellaneous curves.

French Door. A door with rectangular glass panes extending the full length.

French Drain. A trench filled with gravel, often with a perforated pipe on the bottom, to carry off intercepted subterranean drainage.

French Polishing. High-grade wood finish obtained by applying shellac or French varnish with a cloth pad and linseed oil as a lubricant to prevent the pad from sticking.

French Process Zinc Oxide. Zinc oxide pigment made from metallic zinc; sometimes called indirect process; see American Process Zinc Oxide.

French Window. A pair of casement windows that reaches to the floor.

Freon. Trade name for a family of synthetic chemical refrigerants manufactured by E. I. duPont de Nemours & Co., Inc.

Freon-12. Dichlorodifluoromethane.

Frequency. 1. Commonness of occurrence. 2. The rate of recurrence of a vibration, oscillation, or cycle; the number of repetitions in a given time, such as per second; one vibration per second is one hertz.

Frequency Meter. A device used to measure vibrations in cycles per second.

Fresco. The art of painting on fresh plaster with water-based pigments.

Fresh Water. Those waters having a specific conductivity less than a solution containing six thousand (6,000) parts per million of sodium chloride.

Fret. A decorative pattern of interlaced designs.

Fretsaw. A finetoothed saw with its narrow blade held under tension, used for sawing curves.

Fretwork. 1. Ornamental openwork designs in woodwork cut by a fretsaw. 2. In masonry, any ornamental openwork or work in relief.

Friable Asbestos. Any materials that contain greater than one percent asbestos, and which can be crumbled, pulverized, or reduced to powder by hand pressure; this may also include previously non-friable material which becomes broken or damaged by mechanical force.

Friction. 1. The rubbing of one material against another. 2. The tangential surface resistance between two bodies in contact which move or tend to move with respect to each other.

Frictional Soil. A soil, such as sand, in which there is little or no attraction between its particles, and which derives its strength from geometric interlocking of the particles; a non-cohesive soil.

Friction Catch. Any catch which, when engaged in a strike, is held in the engaged position by friction.

Friction Connection. Two or more structural steel members clamped together by high-strength bolts with sufficient force that the loads on the members are transmitted between them by friction along their mating surfaces.

Friction Hinge. A hinge designed to hold a door at any desired degree of opening by means of friction control incorporated in the knuckle of the hinge.

Friction Loss. Loss in efficiency due to friction.

Friction Pile. A pile calculated to carry all of its load by skin friction, neglecting any contribution of direct bearing on its point.

Frieze. 1. A horizontal, often decorative, member of a cornice, set flat against a wall. 2. More broadly, any sculptured or ornamental band on a house, in a room, or on furniture.

Frigidarium. The room containing the cold bath in Roman baths.

Fringe Benefits. Benefits paid for by an employer for an employee in addition to basic wages; includes such benefits as vacations, sick time, health care, retirement, and disability insurance.

Frit. 1. The wholly or partly fused materials of which glass is made. 2. Ground-up glass used as a basis for glaze or enamel.

Fritted Glass. Float glass with crushed glass melted onto its surface, for the purpose of screening out glare, limiting solar gain, and providing various levels of opacity and color.

Fritted Glaze. A glaze in which a part or all of the fluxing constituents are pre-fused.

Froe. A cleaving tool consisting of a wedge-shaped blade mounted at right angles to the blade; used for splitting blocks of wood into shingles or barrel staves.

Frog. 1. A loop on a belt to carry a tool. 2. A mechanical device that allows the train wheels on one track to cross over an intersecting track. 3. See Panel, 8.

Front Door Lock. A lock assembly mounted in a front door.

Front End Loader. A tractor or bulldozer with a bucket which operates from the front of the vehicle.

Front End Loading. The fraudulent practice of a contractor's distorting the schedule of values so that work done early in the contract will have higher values than work done later, for the purpose of obtaining payment before it is earned; also called Unbalancing the Schedule of Values.

Front Money. The cash needed to pay all the property development costs before financing can be put in place.

Front Yard. The space between a building and the front property line.

Frost. 1. A covering of minute ice crystals that form on a cold surface. 2. Frozen soil.

Frost Back. Condition in which liquid refrigerant flows from the evaporator into the suction line; usually indicated by sweating or frosting of the suction line.

Frost Control, Automatic. See Automatic Frost Control.

Frost Control, Manual. See Manual Frost Control.

Frost Control Semiautomatic. See Semiautomatic Frost Control.

Frost Free Refrigerator. Refrigerated cabinet which operates with an automatic defrost during each cycle.

Frosting Type Evaporator. Refrigerating system which maintains the evaporator at frosting temperatures during all phases of cycle.

Frost Line. The greatest depth to which ground may be expected to freeze.

Frostline. The depth of frost penetration in soil; this depth varies in different parts of the country; footings should be placed below this depth to prevent movement.

Frost Proof Tile. Tile produced for use where freezing and thawing conditions occur.

Frothing. Foaming.

Frozen. 1. Water in its solid state. 2. Seized, as in machine parts, due to lack of lubrication; the term Freeze-Up is often applied to this situation.

Frozen Food Case. A storage unit that freezes or maintains foods at freezing temperatures.

FRP. Fiberglass Reinforced Plastic.

Fryer, Kitchen. See Kitchen Fryer.

FS. Federal Specifications.

FTI. Facing Tile Institute.

Fuel. A substance burned to provide heat.

Fuel Oil. 1. Any oil that is used for fuel in engines or furnaces. 2. Kerosene or any hydrocarbon oil having a flash point not less than 100° F. (38° C.).

Fugitive Colors. Colors which are not permanent; subject to fading.

Fulcrum. A fixed point of support about which a lever is pivoted.

Full-Cell Process. A process for impregnating wood with preservatives or chemicals; in this process a vacuum is formed to remove air from the wood before adding the preservative.

Full-Depth Asphalt Pavement. The term Full-Depth, registered by The Asphalt Institute with the U. S. Patent Office, certifies that the pavement is one in which asphalt mixtures are employed for all courses above the subgrade or improved subgrade; a Full- Depth asphalt pavement is laid directly on the prepared subgrade.

Fuller's Earth. A type of clay mineral that lacks plasticity and is used as an adsorbent and as a filter.

Full Fillet Weld. A fillet weld whose size is equal to the thickness of the thinner member joined.

Full Floating. Mechanism construction in which a shaft is free to turn in all the parts in which it is inserted.

Full Frame. In cabinet construction, amortised-and-tenoned frame, in which every joint is mortised and tenoned.

Full Header. A brick course composed of all headers.

Full Load Torque. In an electric motor, the maximum torque delivered without overheating.

Full Mopping. In roofing, a mopping layer that completely coats the surface with hot bitumen.

Fully Adhered. A roof membrane that is fully mopped to the substrate.

Fume Hood. A ventilated laboratory enclosure for disposing of fumes.

Functional. Designed to be practical; utilitarian.

Functionalism. A philosophy of design holding that form should be adapted to use, material, and structure.

Functional Obsolescence. The loss in value of real property due to outmoded features or poor design.

Fundamental Period. The time it will take for an element or structure to stop moving and return to its original vertical position when a horizontal force is exerted on it; for example, when a horizontal force is exerted on a building causing displacement.

Fungi. See Fungus.

Fungicide. A substance that destroys fungi or inhibits their growth.

Fungi Resistance. The ability of any material to withstand fungi growth and their metabolic products under normal conditions of service.

Fungus (pl. Fungi). Any of a major group of saprophytic and parasitic lower plants that lack chlorophyll and include molds, rusts, mildews, smuts, mushrooms, and yeasts.

Funnel. A utensil consisting of a tube at the small end of a hollow cone; used for pouring liquids into a container with a small opening.

Funny Papers. A disparaging term used on the jobsite in reference to the construction drawings and specifications

Furan Grout. A two-part grout system of furan resin and furan hardener used for filling joints between quarry tile and pavers where chemical resistant properties are required; also called Furan Resin Grout.

Furan Mortar. A two-part mortar system of furan resin and furan hardener used for bonding tile to back-up material where chemical resistance of floors is important; also called Furan Resin Mortar.

Furan Plastics. Plastics based on resins in which the furan ring is an integral part of the polymer chain, made by the polymerization or polycondensation of furfural, furfuryl alcohol, or other compounds containing a furan ring, or by the reaction of these furan compounds with other compounds, the furan being in greater amount by weight.

Furan Resin Grout. See Furan Grout.

Furan Resin Mortar. See Furan Mortar.

Furan Resin. See Furan Plastics.

Furnace. A unit which draws in cool air from an occupied space and passes the air through a heating chamber, combustion or electric, and then is returned to the occupied space; a heat system using air as the distribution fluid.

Furnace, Central Warm Air. Self-contained appliance designed to supply heated air through ducts to spaces remote from or adjacent to the appliance location.

Furnish. 1. To provide what is needed. 2. To supply furniture. 3. The raw materials needed for making paper pulp.

Furnishings. Articles, especially furniture, found in the interior of a structure, generally to increase comfort or utility.

Furred Ceiling. A ceiling having spacer elements, usually furring channels, round rods, or wood strips, interposed between it and the supporting structure above.

Furring. 1. Wood or metal strips used to build out a surface such as a studded wall. 2. Narrow strips fastened to the walls and ceilings to form a straight surface upon which to lay the lath or other finish.

Furring, Ceiling. See Ceiling Furring.

Furring, Channel. See Channel Furring.

Furring Channels. 3/4 inch cold or hot rolled steel channels used in plaster base construction.

Furring, Metal. See Metal Furring.

Furring, Wood. See Wood Furring.

Furrowing. To strike a V shaped trough in a bed of mortar.

Fuse. An electrical safety device consisting of strip of fusible metal in circuit which melts when circuit is overloaded.

Fused Reducer. A pipe coupling with a larger size at one end that the other and is attached to a length of pipe by welding.

Fusel Oil. Oily liquid produced in small quantities when ethyl alcohol is produced by fermentation of grain.

Fusible Link. A safety device consisting of a metal of low melting point inserted into a machine, boiler, or linkage that will melt if the temperature rises above a certain point thereby closing or shutting off the appliance.

Fusible Plug. Plug or fitting made with a metal of a known low-melting temperature; used as safety device to release pressure.

Fusible Switch. An electric switch that has a fusible link in the wiring for an electric circuit to break the circuit by melting in the event of an overload.

Fusion. 1. The liquid or plastic state of a material caused by heat. 2. In welding, the process of melting to form a union; usually the result of interaction of two or more materials.

Fuzzing. A temporary condition on new carpet consisting of irregular fuzzing appearance caused by slack yarn twist, fibers snagging, or breaking of yarn; can be remedied by spot shearing.

G

GAAP. Generally Accepted Accounting Principles.

Gabion. A metal or wire cage filled with ballast or stone, used in large scale retaining walls.

Gable. The triangular wall beneath the end of a gable roof; the vertical triangular end of a building from the eaves to the apex of the roof.

Gable End Rafter. The last rafter system installed at the gable end of a building.

Gable End. An end wall having a gable.

Gable Louver. A louver system installed at the gable end of a building.

Gable Roof. A roof consisting of two opposite sloping planes that intersect at a level ridge.

Gable Truss. The truss installed at the gable end of a building.

Gable Vent. A screened of louvered, opening in a gable, used for venting from an attic.

Gaffer. A lighting electrician on a motion picture or television production unit.

Gage. A tool used by carpenters to strike a line parallel to the edge of a board.

Gain. 1. A notch or mortise, as in a beam or a wall, for a joist, girder or similar member. 2. The notch to set a hinge in a door or frame.

Gallery. 1. A roofed outdoor promenade or balcony. 2. A room or building for displaying art or artifacts.

Galvanic Action. See Galvanic Corrosion.

Galvanic Corrosion. Corrosion in a metal caused by a galvanic cell created by dissimilar metals in close proximity accompanied by an electrolyte, often atmospheric moisture; also called Galvanic Action.

Galvanized. Zinc plated for corrosion protection achieved by hot dipping into molten zinc or by electrolysis.

Galvanized Accessory. A building product such as a door, window, skylight, handrail, grating, or ventilator, that has been galvanized.

Galvanized Iron. Zinc coated iron or steel.

Galvanized Mesh. Mesh screening that has been galvanized; used as wire lath, reinforcing, or fencing.

Galvanized Metal Joist Hanger. Formed steel component used to support end of load bearing joists and transmit loads to another joist or beam, galvanized.

Galvanized Pipe. A steel pipe galvanized to prevent rusting.

Galvanizing Repair. Repair of damaged galvanized surfaces by application of zinc rich paint.

Gambrel Roof. A type of gable roof where the roof planes on each side of the ridge each have two pitches, the lower being steeper and longer than the upper; also called a Curb Roof.

Gang Box. Electrical rough-in box constructed of metal or hard plastic, to provide for two or more outlets or switches.

Gang Nail. A sheet metal plate that has numerous sharp tabs that act as nails to hold the sheet metal plate over a joint in a light wood truss.

Gangway. 1. A passageway. 2. A temporary passage over sometimes cleated planks on a construction site.

Gap. 1. An empty space between two objects. 2. A break in continuity.

Gap-Graded Aggregate. Aggregate containing particles of both large and small sizes, in which particles of certain intermediate sizes are wholly or substantially absent.

Garage. A room or building for storage of automobiles.

Garage Door Opener. An electrically driven mechanical device to raise and lower a garage door, usually activated by a radio control in the car or by a wall switch.

Garbage Disposal Unit. An electrically powered device, usually located in a kitchen sink, to grind household garbage and discharge it to the sewage system.

Garbage Handling. The act or process of the transport and discharge of waste materials.

Garden Wall Bond. A name that is given to any brick bond that is especially suitable for a wall two tiers thick, consisting of one header to three stretchers in every course.

Garnet Paper. Abrasive which is reddish in color, hard and sharp; comes from same source as semi-precious jewel by that name; more expensive than flint paper but lasts longer.

Garret. An attic room or attic.

Gas. 1. Any airlike substance which moves freely to fill the available space regardless of the quantity. 2. A combustible gaseous mixture; fuel gas. 3. Vapor; one of the three states of matter; compare with Liquid and Solid.

Gas Checking. Fine checking; wrinkling, frosting under certain drying conditions; said to be caused by rapid oxygen absorption or by impurities in the air.

Gas Cock. A fuel gas valve.

Gas Concrete. See Cellular Concrete.

Gas Furnace. A heating system that burns gas to produce heat.

Gasholder. A large cylindrical tank for storing gas under pressure.

Gasket. Sealing material at a crack; a sheet or ring of rubber, vinyl, or similar material shaped to seal the junction of metal surfaces.

Gasket, Foam. See Foam Gasket.

Gas Log. A hollow perforated imitation log used as a gas burner in a fireplace.

Gas Metal Arc Welding (GMAW). An arc welding process wherein coalescence is produced by heating with an arc between a continuous filler metal (consumable) electrode and the work. Shielding is obtained entirely from an externally supplied gas, or gas mixture. Some methods of this process are called MIG or CO welding.

Gas Meter. A mechanical device for measuring and recording the volume of gas passing a given point.

Gas, Noncondensable. See Noncondensable Gas.

Gas Pocket. A weld cavity caused by entrapped gas.

Gas Pump. A mechanical device to move gas or gasoline from one location to another.

Gas-Tight Manhole Cover and Frame. Air tight metal cover and frame over vertical access shaft from grade level down to underground utility or chamber.

Gas Tungsten Arc Welding (GTAW). An arc-welding process wherein coalescence is produced by heating with an arc between a single tungsten (nonconsumable) electrode and the work; shielding is obtained from a gas or gas mixture; pressure may or may not be used and filler metal may or may not be used; also called TIG welding.

Gas Valve. Device in a pipeline for starting, stopping or regulating flow of gas.

Gas Welding. A welding process wherein coalescence is produced by heating with a gas flame or flames, with or without the application of pressure, and with or without the use of filler metal.

Gasworks. A place where gas is manufactured and processed.

Gate, Fence. See Fence Gate.

Gate Post. The vertical member in a fence system upon which the gate hinges are mounted.

Gate, Swing. See Swing Gate.

Gate Valve. A valve utilizing a wedge-shaped gate, which allows fluid flow when the gate is lifted from the seat.

Gateway. 1. A gate. 2. An opening for a gate.

Gauge. 1. Measurement according to some standard system. 2. The thickness of sheet metal. 3. An instrument for measuring the size or quantity. 4. The distance between tufts across the width of knitted and tufted carpets, expressed in fractions of an inch.

Gauge, Compound. See Compound Gauge.

Gauged Brick. 1.Brick which have been sorted or otherwise produced to accurate dimensions. 2. A brick that has been rubbed on an abrasive stone to reduce it to a trapezoidal shape for use in an arch.

Gauge, High Pressure. See High Pressure Gauge.

Gauge, Low Pressure. See Low Pressure Gauge.

Gauge, Manifold. See Manifold Gauge.

Gauge Port. Opening or connection provided for a service technician to install a gauge.

Gauge Pressure. The pressure indicated on a pressure gauge, measured in pounds per square inch, and indicated as PSIG; compare with Absolute Pressure, indicated as PSIA.

Gauge, Vacuum. See Vacuum Gauge.

Gauging. 1. Another cementitious material (usually calcined gypsum, Keene's cement, or portland cement) added to lime putty to provide and control set. 2. The act of adding gauging material.

Gauging Plaster. A gypsum plaster formulated for use in combination with finish lime in finish coat plaster.

Gauging Trowel. The gauging trowel is larger than the pointing trowel but smaller than the buttering trowel.

GB. Gigabyte. 1,024 megabytes.

Gel. A semi-solid colloidal suspension or jell; a solid dispersed in a liquid; a jelly-like substance.

Gelatin. A product of the packing house; a glutinous material obtained from animal tissues by boiling; which can be cast into a semi-rigid mould; on account of its flexibility, it is particularly adaptable to moulds containing undercuts.

General and Administrative Expenses. Shown in the income statement as all items of expense of a general nature which cannot be specifically attributed to individual construction projects.

General Conditions. A written document, supplementing the specifications, which indicates and defines areas of the project relating to other than specific building trades.

General Contract. The agreement covering the work of general construction done by a prime contractor.

General Contractor. A prime contractor for general construction.

General Drawing. 1. A drawing showing elevations, plans and cross sections of the structure and the main dimensions. 2. A drawing showing the borings for substructure.

General Journal. The record used for recording business transactions not recorded in specific journals.

General Ledger. The record containing the summarization of all transactions of a business.

General Partner. A member of a partnership with unlimited management authority and liability.

General Partnership. Embodies the financial and personal resources of two or more persons who share in the owning and running of the business.

General Requirements. See Job Requirements.

Generator. A machine by which mechanical energy is changed into electrical energy.

Gentrification. The purchase and rehabilitation by the middle-classes of run-down properties.

Geodesic. The shortest line between two points on a given surface.

Geodesic Dome. A dome constructed according to the patents of R. Buckminster Fuller, in which the pattern of surface divisions is always a function of an entire sphere; commonly constructed with prefabricated structural triangles linked together in a mosaic to create the domed shape.

Geoid. The sphere that defines the earth at the location of mean sea level.

Geometry. A branch of mathematics that deals with the measurement, properties, and relationships of points, lines, surfaces, and solids.

Geotechnical Engineer. An engineer who specializes in rock and soil mechanics, groundwater, and foundations.

Geotechnical Investigation. The process of boring, sampling, and testing the soil at various depths to enable the geotechnical engineer to discover and analyze characteristics of the soil; also called Subsurface Investigation, Foundation Investigation, Soil Investigation, or Soil Test; the test results and engineer's recommendations are explained and summarized in a Soil Report.

Geotextile. Synthetic fabrics used to separate backfill materials for proper drainage; used in high retaining walls and landscape design.

Gesso. A plaster surface composed of gypsum plaster, whiting, and glue, used as a base for decorative painting.

GFI Breaker. Ground Fault Interrupter Breaker; supplies power, as any breaker does, but also monitors the amount of incoming and outgoing current; whenever the entering current does not equal the leaving current, indicating current leakage, the GFI instantly opens the circuit; a faster overcurrent protection device than either a fuse or circuit breaker.

GFRC. Glass-Fiber-Reinforced Concrete.

Ghosting. In painting, patches of lighter color showing in dry coat; a coating with a skippy appearance.

GI. Galvanized Iron.

Gild. To apply a thin covering of gold.

Gilding. Process of obtaining a finish by using metal leaf.

Gillmore Needle. A device used in determining time of setting of hydraulic cement.

Gilsonite. A black, coal-like substance obtained from mines and used in the manufacture of black asphaltum varnish.

Gin Pole. A simple derrick consisting of a pole held nearly vertical by guy cables, a pulley at the top of the pole, and a hoisting rope.

Girder. A beam that supports other beams; a very large beam, especially one that is built up from smaller elements; a timber beam used to support wall beams or joists.

Girt. A beam that supports wall cladding between columns.

Gland. A device for preventing leakage of fluid past a joint in machinery, as at a bearing.

Glare. Excessive brightness in the field of view.

Glass. A hard brittle usually transparent, translucent, or shiny substance, made by fusing sand with soda and lime.

Glass Bead. A narrow strip of plastic, metal, or wood used to hold glass in a sash; removable trim that holds glass in place.

Glass Block. A hollow masonry unit made of glass.

Glass Cutter. A tool for cutting glass.

Glass Door. A metal or wood framed door manufactured with heat-strengthened or tempered glass.

Glass Felt. A felt sheet in which glass fibers are bonded into the felt sheet with resin; they are suitable for impregnation and coating; used in the manufacture and coating of bituminous waterproofing materials, roof membranes, and shingles.

Glass Fiber. Glass in the form of fine fibers used in fabrics.

Glass-Fiber-Reinforced Concrete (GFRC). Concrete with a strengthening admixture of short alkali-resistant glass fiber.

Glass Mat. A thin mat of glass fibers with or without a binder.

Glass Mosaic Tiles. Tiles made of glass, usually in sizes not over two inches square and 1/4 inch thick, mounted on sheets of paper, usually twelve inches square.

Glass Mullion System. A method of constructing a large glazed area by stiffening the sheets of glass with perpendicular glass ribs.

Glass Pipe. Glass and glass-lined pipe used in process piping and in laboratories.

Glassware Washer. An apparatus for the cleaning of items made of glass.

Glass, Wire. See Wire Glass.

Glass Wool. Glass in the form of fine fibers used in insulation.

Glaze. A ceramic coating matured to the glassy state on a formed ceramic article; also refers to the material or mixture from which the coating is made.

Glaze Coat. 1. The top layer of asphalt used on a smooth-surfaced built-up roof assembly. 2. A thin protective layer or coating of bitumen that is applied to the lower plies or the top ply of a built-up membrane when the top pouring and aggregate surfacing are delayed.

Glazed Block. Concrete blocks with a surface produced by fusing it with a glazing material.

Glazed Brick. Brick or tile with a surface produced by fusing it with a glazing material.

Glazed Bullnose. A ceramic trim tile with a convex radius on one edge that has been given a glassy or glossy surface.

Glazed Ceramic Mosaic Tile. A glazed tile with a body that is suitable for interior use and which is usually nonvitreous and is not required to withstand excessive impact or be subject to freezing.

Glaze Fit. The stress relationship between the glaze and body of a fired ceramic product.

Glazed Floor. A floor covering that has had an application of a nearly transparent coating that enhances and protects the coat underneath.

Glazed Interior Tile. A glazed tile with a body that is suitable for interior use and which is usually non-vitreous, and is not required or expected to withstand excessive impact or be subject to freezing and thawing conditions.

Glazed Paver Tile. See Paver.

Glazed Quarry Tile. See Quarry Tile.

Glazed Structural Unit, GSU. A solid or hollow unit with a surface of applied smooth glossy nature, e.g. a tile with a fired glaze finish.

Glazed Tile. Tile with a fused impervious facial finish composed of ceramic materials, fused into the body of the tile which may be a non-vitreous, semi-vitreous, vitreous, or impervious body; the glazed surface may be clear, white, or colored.

Glazed Tile, Extra Duty Glaze. Tile with a durable glaze that is suitable for floors and all other surfaces.

Glazed Wall Tile. A wall tile with an impervious, glossy finish.

Glazier. One whose trade is setting glass.

Glazier's point. A small triangular sheetmetal nail to keep glass in place in a wood sash before puttying.

Glazing. 1. The trade of installing glass; the trade practiced by glaziers. 2. A transparent or translucent color applied to modify the effect of a painted surface. 3. In plastering, a condition created by the fines of a machine-dash texture plaster traveling to the surface and producing a flattened texture and shine or discoloration; this may be caused by the basecoat being too wet or the acoustical mortar being too moist; glazing occurs in hand application when mortar being worked is excessively wet.

Glazing Accessory. Implements or devices needed in the glazing trade.

Glazing Bead. A strip of metal or wood used to hold a sheet of glass in place.

Glazing Bead. A strip of wood or metal for holding a pane of glass in its frame or sash.

Glazing Compound. Any of a number of types of mastic or putty used to bed small lights of glass in a frame.

Glitter. A reflective material such as glass, diamond dust, or small pieces of variously colored aluminum foil projected into the surface of wet plaster or paint as a decorative treatment.

Globe Valve. A spherically shaped valve body which controls the flow of water with a compression disc which is opened and closed by means of a stem and mates with a ground seat to stop water flow.

Globular Transfer. A mode of metal transfer in gas metal-arc welding in which the consumable electrode is transferred across the arc in large droplets.

Gloss. A shiny, lustrous finish which reflects light. The term also refers to paint or enamel that dries to a high sheen or luster, usually with a hard, smooth coat.

Gloss Enamel. A finishing material made of varnish and pigments; such an enamel forms a hard coating with a smooth surface and high gloss.

Gloss Oil. A varnish composed primarily of limed rosin and petroleum thinner.

Gloss Retention. Ability to retain original sheen.

Glost Fire. The process of kiln-firing bisque ware to which glaze has been applied.

Glovebag. A polyethylene or polyvinyl chloride bag-like enclosure with integral gloves affixed around an asbestos-containing source so that the material may be removed while minimizing release of airborne fibers to the surrounding atmosphere.

Gloves. Protective covering for the hands to eliminate injury and improve grip.

Gluability. The ease or difficulty in bonding a material with adhesive; bondability.

Glue. Adhesive; commonly used in joining wood parts.

Glue Block. A small piece of wood used to strengthen and support two pieces of wood joined at an angle; see Corner Block.

Glue Built-Up Members. Structural elements, the sections of which are composed of built-up lumber, wood structural panels, or wood structural panels in combination with lumber; all parts bonded together with adhesives.

Glue Gun. A hand tool for the application of bulk or cartridge-type adhesives.

Glue Joint. A joint held together with glue.

Glulam. Glue laminated timber.

Glue Laminated Member. Assemblies of wood laminations bonded together with adhesive so that longitudinal grain of laminations are approximately parallel; Architectural Grade commonly has a better degree of external finish and would be exposed in the finished building; Industrial grade would not be finished to the same degree as Architectural Grade.

Glue Laminated Timber. A timber made up of a large number of small strips of wood glued together.

Glycerine. Glycerol.

Glycerol. A sweet viscous liquid alcohol, almost colorless and odorless; used extensively in the production of alkyd resins and ester gum.

Glycol. Ethylene glycol, a thick chemical alcohol, used as an antifreeze.

Glyph. 1. A sculptured character or symbol. 2. A vertical groove as on a Doric frieze.

GMP. Guaranteed Maximum Price.

Gneiss. A usually coarse grained metamorphic rock foliated by mineral layers, principally of feldspar, quartz, and ferromagnesian minerals.

Gnomon. 1. The rod, pin, or style on a sundial that casts its shadow indicating the time. 2. The L shaped remainder of a parallelogram after the removal of a smaller similar parallelogram containing one of its corners.

Goggles. Protective covering for the eyes.

Going Concern Concept. The accounting assumption that a business will continue to operate for an indefinite period.

Gold. A yellow malleable ductile high density precious metallic element resistant to chemical reaction, occurring naturally in quartz veins and gravel.

Gold Leaf. Gold beaten into a very thin sheet, ordinarily 4 to 5 millionths of an inch in thickness.

Gold Paint. Mixture of bronze powder and bronzing liquid.

Golf Shelter. A structure placed on a golf course to be used for the safety of golfers in inclement weather.

Gondola. 1. A railroad car with a flat bottom, no top, and fixed sides for shipping heavy construction materials and equipment. 2. An island of shelves to display goods in a supermarket.

Goodwill. The value of a business over the book value or agreed value of its net assets based on its experience and reputation.

Goose Neck. A blacksmith's tool used to handle hot metals.

Gothic. An architectural style of Western Europe of the 12-16th centuries featuring pointed arches, vertical piers, buttresses, and vaults.

Gothic Arch. A pointed arch, with a joint at the apex rather than a keystone.

Gouge. 1. A chisel with a concave blade. 2. To overcharge.

Grab Bar. Metal or plastic bar attached to a bathroom wall, above a bathtub, near a toilet, or in a shower, to be used as a hand hold. See Flash Set.

Graben. A long, narrow trough bounded by one or more parallel normal faults; also called a Rift Valley.

Grad. A metric unit of circular measurement; 400 grads equals a full circle equals 360 degrees.

Gradall. A hydraulic, wheel-mounted backhoe often used with a wide bucket for dressing earth slopes.

Gradation. The sizing of granular materials. for concrete materials, usually expressed in terms of cumulative percentages larger or smaller than each of a series of sieve openings or the percentages between certain ranges of sieve openings.

Grade. 1. The classification of lumber in regard to strength and utility in accordance with the rules of an approved lumber grading agency. 2. A gradient. 3. A predetermined degree of slope that a finished floor or ramped surface should have. 4. The horizontal ground level of a building or structure. 5. To level off to a smooth horizontal or sloping surface. 6. The slope or pitch, known as the fall, usually expressed in drainage piping as a fraction of an inch per foot. 7. The elevation of a real or planned surface or structure.

Grade Beam Formwork. The system of support for freshly poured or placed concrete in a reinforced grade beam.

Grade Beam Reinforcing. The placing of metal or steel bars in a concrete grade beam.

Grade Beam. A reinforced concrete beam that transmits the load from a bearing wall into spaced foundations such as pile caps or caissons.

Graded Sand. A mixture of sand in which the granules consist of fine, medium, and coarse sizes.

Grade Mark. The stamp on lumber indicating its grade according to the rules of the lumber grading bureau that performed the inspection.

Grade MW Brick. A grade of brick intended to be used for exterior wall surfaces where water permeability is not expected and only slight frost action; MW means moderate weather.

Grade NW Brick. A grade of brick intended to be used for back up withes or interior exposure; NW means no weather.

Grade Stake. A stake indicating the amount of cut or fill required to bring the ground to a specified level.

Grade SW Brick. A grade of brick intended to be used for exterior wall surfaces and where water permeability and frost action may be expected; SW means severe weather.

Gradient. 1. A stretch of road or railway that slopes from the horizontal. 2. The amount of such a slope. 3. The rate of rise or fall of temperature or pressure in passing from one region to another.

Grading. See Grade.

Gradiometer. A geophysical instrument that measures the vertical component of the earth's magnetic field; a very sensitive instrument, it can detect minor variations due to buried features.

Grain. 1. The direction, size, arrangement, appearance, or quality of the fibers in wood or lumber. 2. The weight unit equal to 1/7000 pound; used in measuring atmospheric moisture content.

Grain Alcohol. Ethyl alcohol made from grain by distillation.

Graining. Simulating the grain of wood by using paint.

Grainness. Roughness of a protective film resembling grains of sand.

Grain Raising. Causing short fibers on surface of bare wood to stand up by applying water; liquids that do not raise the grain are known as non-grain-raising.

Gram. A metric unit of mass and weight equal to 1/1000 of a kilogram and nearly equal to one cubic centimeter of water at its maximum density.

Gram Calorie. See Calorie, 1.

Grand Master Key. A key that operates locks in several groups, each of which has its own master key.

Grandstand Seat. A seat in a stand for spectators at a racecourse or stadium.

Grandstand. A stand or bleachers for spectators, usually for a sporting event.

Granite. Igneous rock with visible crystals of quartz and feldspar.

Granite Block. A masonry unit consisting of a very hard natural igneous rock used for its firmness and endurance.

Granite Curb. Street curbing that is manufactured from granite blocks.

Granite Paver. .A type of igneous stone that is harder and more durable than regular bricks, used for driveways and patios for use with or without mortar.

Granite Veneer. A masonry facing attached to a main structure, made from granite.

Granolithic Topping. A covering layer consisting of an artificial stone of crushed granite and cement.

Grant Deed. A deed by which ownership in real property is transferred.

Grantee. A person receiving an interest in property.

Grantor. A person selling, granting, or giving up an interest in property.

Granular Insulation. Insulating material made of granular vermiculite or perlite or beaded styrene, placed in hollow, usually masonry, construction.

Granular Waterproofing. Waterproofing material made of granular vermiculite or perlite or beaded styrene.

Granules. The size-graded mineral particles that are imbedded into the top surface of factory made mineral-surfaced asphalt shingles and cap sheet.

Graph. A diagram showing the relation between variable quantities, usually two variables, each measured along one of a pair of axes at right angles.

Graphic. Of or relating to the visual or descriptive arts, like writing, drawing, or photography.

Graphics. The products of the graphic arts.

Graphite. Black pigment consisting mostly of carbon obtained from natural deposits or produced from coke in an electric furnace.

Grass. Green low growing vegetation widely used as ground cover; see Lawn.

Grass Cloth Tile. Tile made from woven vegetable fibers that are laminated.

Grass Cloth. A wall covering manufactured from vegetable fibers, woven and laminated to paper for backing.

Grate Cutting. The act or process of snipping away a grate with hand tools, or cutting away a grate by acetylene torch or power tools.

Grate, Metal. See Metal Grating.

Grate Painting. The act or process of applying paint to seal, prevent rust, or add color to a grating.

Grating. Open grid of metal bars structurally formed.

Grating Frame. Frame, usually metal, to contain floor grating and provide means to anchor to floor construction.

Grating, Metal. See Metal Grating.

Grating Tread. Horizontal surface of step made of steel grating.

Gravel. Small rock particles resulting from natural disintegration and weathering such as river gravel; mechanically crushed stone.

Gravel Fill. Gravel layered at the bottom of an excavation to insure drainage of water.

Gravel Mulch. A protective covering of rock particles, used on the ground to reduce evaporation, prevent erosion, and control weeds.

Gravel Roof. A roof composed of layers of roofing felt for waterproofing, then sealed with tar or pitch and covered with a layer of gravel to assist in protection from wear and the sun.

Gravel Stop. A metal flange or strip with a vertical lip placed around a built-up roof to prevent loose gravel from falling off the roof; it is also used to present a finished edge detail in built-up roofing.

Gravity. 1. The force that attracts a body to the center of the earth. 2. The quality of having weight.

Gravity Furnace. An appliance, with a heat exchanger, that supplies hot air that is distributed by gravity.

Gravity Ventilator. A device installed in an opening in a room or building which is activated by air passing through to remove stale air and replace it with fresh air.

Gravity Wall. A retaining wall which depends solely on its weight to resist lateral forces of retained earth.

Gray Blast Cleaning. Commercial blast.

Grayish. Lacking in intensity of color.

Gray Paper. The unsized, uncalendered paper used on the back side of regular gypsum board products and as the face and back paper on backing board products.

Grease Trap. A device to trap and retain the grease content of wastewater and sewage.

Great Calorie. See Calorie, 2.

Green Architecture. Design that embraces energy conservation and minimal use of fossil fuels and non-renewable resources.

Green Board. A gypsum board having a tinted face paper usually light green (or blue) to distinguish special board types.

Green Brickwork. Brickwork in which the mortar has not yet set.

Green Concrete. Concrete which has set but which has not appreciably cured or hardened to its design strength.

Green Design. See Green Architecture.

Greenfield Site. A building site that has never before been built upon; compare with Brownfield Site.

Greenhouse. A glassed enclosure used for cultivation or protection of tender plants.

Greenhouse Effect. The trapping of the sun's warmth in the lower atmosphere of the earth caused by an increase in carbon dioxide, which is more transparent to solar radiation from the earth.

Green Lake. A pigment; a mixture of Prussian blue and yellow lake, sold under various trade names.

Green Masonry. In masonry, a molded clay unit before it has been burned in preparation for building purposes; an uncured concrete masonry unit.

Green Plaster. Newly applied plaster that has not dried.

Green Room. A room in a theater for actors who are off-stage.

Green Wood. Freshly sawed or undried wood; twenty percent moisture or more.

Gremlin. A mischievous creature said to be responsible for malfunctions of machinery and equipment.

Grid. 1. A network of uniformly spaced horizontal and perpendicular lines, as for locating and referring to points on a drawing. 2. A set of surveyor's closely spaced reference lines laid out at right angles, with elevations taken at line intersections.

Griddle, Kitchen. A flat surface upon which food is cooked by dry heat.

Grillage. A framework composed of main runner channels and furring channels to support ceilings.

Grille. Component of the ventilation system that promotes air circulation in the occupied space by providing a means to return air; a metal screen or grating that allows for the circulation of air.

Grin. Condition where the backing shows through sparsely spaced pile tufts; carpets may be grinned (bent back) deliberately to reveal the carpet construction.

Grinder, Electric. See Electric Grinder.

Grind Gauge, Hegeman. See Hegeman Grind Gauge.

Grit. An abrasive obtained from slag and other materials.

Grommet. Plastic, metal or rubber doughnut-shaped protector which lines holes where wires or tubing pass through wiring boxes or panels.

Groove. 1. The opening provided for a groove weld. 2. See Dado. 3. See Housing. 4. See Tongue and Groove.

Groovc Weld. A weld made in the groove between two members to be joined; the standard types of groove welds are. Square, Single-Vee, Single-Bevel, Single-U, Single-J, Double-Vee, Double-Bevel, Double-U, and Double-J.

Gross. 1. An overall total exclusive of deductions. 2. A quantity of 12 dozen; 144.

Gross Building Area. The total building area, without deductions, expressed in square feet.

Gross Cross Sectional Area. Area measured by overall dimension, including voids as area.

Gross Errors. Errors that are easily detected as they are large in proportion to the context in which they arise.

Gross Income. The total income produced by a real estate investment, without deduction for loan repayment or expenses.

Gross Leasable Area. The total area of a building that is leasable to tenants; the gross building area less the unrentable areas.

Gross Multiplier. A rule-of-thumb method of appraising the value of income property by multiplying the gross annual income by a multiplier; this is an unreliable method, used primarily on residential property.

Gross Profit. The excess of net sales over the cost of goods sold.

Ground. 1. An electrical connection with the earth. 2. A conducting connection whether intentional or accidental, between an electrical circuit or equipment and earth, or to some conducting body which serves in place of the earth. 3.A strip of wood assisting the plasterer in making a straight wall and in giving a place to which the finish trim of the room may be nailed. 4. A strip attached to a wall or ceiling to establish the level to which plaster should be applied.

Ground Acceleration. Acceleration of the ground due to earthquake forces.

Ground Bus. In the main electrical service panel, where the neutral service wire attaches and is linked to the earth by the ground wire.

Ground Coat. In painting, the coating material which is applied before the graining colors or glazing coat.

Ground Coil. Heat exchanger buried in the ground; may be used either as an evaporator or as a condenser.

Ground Color. The background color against which the top colors create the pattern or figure in the design.

Ground Cover. Low lying planting materials that cover the ground, such as grass and ivy. 2. See Soil Cover.

Ground Displacement. The distance which ground moves from its original position during an earthquake.

Grounded. Connected to the electrical circuit grounding conductor where one exists; or in other cases, connected to a grounding electrode meeting the required characteristics of the installation.

Grounded Conductor. A system or circuit conductor which is intentionally grounded.

Ground Failure. A situation in which the ground does not hold together such as during a landslide, mud flow, or liquefaction.

Ground Fault Circuit Interrupter (GFCI). Supplies power, as any receptacle does, but also monitors the amount of incoming and outgoing current; whenever the entering current does not equal the leaving current, indicating current leakage, the GFCI instantly opens the circuit. A faster overcurrent protection device than either a fuse or circuit breaker.

Ground Hydrant. A water hydrant for the use in fighting fires, installed in the ground.

Grounding. 1. The act or process of making an electrical connection with the earth. 2. A large conduction body, as the earth, used as a common return for an electric circuit.

Grounding Clip. A metal clipping device which is used to fasten a grounding wire to a pipe or fixture.

Grounding Conductor. A conductor that is used to ground electrical equipment or a grounded circuit to a grounding electrode.

Grounding Conductor, Common Main. In a grounded system, the conductor that connects both the circuit grounded conductor and the equipment grounding conductor to the grounding electrode.

Grounding Conductor, Equipment. A conductor used to connect the equipment being grounded to the service equipment enclosure.

Grounding Conductor, Main. In an ungrounded system, the conductor connecting the equipment grounding conductor at the service to the grounding electrode.

Grounding Locknut. A locknut serving as a connecting body between electrical equipment and a ground wire.

Grounding Plug. A type of electrical plug which grounds a device.

Ground Lease. A lease of the land.

Ground Movement. A general term; includes all aspects of motion. acceleration, velocity, displacement.

Ground Paper. Wallpaper coated with an overall background color.

Ground Rent. Rent consideration under a ground lease.

Ground Rod. A rod acting as a connecting body between electrical equipment and the ground.

Groundscraper. A wide spreading building of offices or apartments; compare with Skyscraper.

Ground Short Circuit. Fault in an electrical circuit allowing electricity to flow into the metal parts of a mechanism.

Ground Water. Water in the subsoil.

Ground Wire. 1. Electrical wire which will safely conduct electricity from a structure into the ground. 2. A wire attached to dissipate electrostatic charge in airless spraying.

Grout. A rich or strong cementitious or chemically setting mix used for filling masonry or tile joints and voids; a mixture of portland cement, aggregates, and water, which can be poured or pumped into cavities in concrete or masonry; also used for filling hollow metal door frames.

Grout-Aid. A proprietary admix to reduce the shrinkage of grout, as placed by the high-lift grouting method.

Grout, Block. See Block Grout.

Grout Bond. The adhesion to, and the interlocking of grout with the masonry units and the reinforcement.

Grouted Hollow-Unit Masonry. That form of grouted masonry construction in which certain designated cells of hollow units are continuously filled with grout.

Grouted Multi-Wythe Masonry. That form of grouted masonry construction in which the space between the wythes is solidly or periodically filled with grout.

Grout, Epoxy. See Epoxy Grout.

Grouting. Process of filling tile joints, masonry block or cells, or any masonry type product with grout.

Grout Lift. An increment of height that grout is poured.

Grout, Masonry. See Masonry Grout.

Grout Mix. The amount of each material specified, portland cement, fine aggregate, admixture, and water.

Grout Pour. The total height of masonry wall to be grouted prior to the erection of additional masonry. A grout pour will consist of one or more grout lifts.

Grout Saw. A saw-toothed carbide steel blade mounted on a wooden handle, used to remove old tile grout, also used in patching work; care should be used, as adjacent tiles can be easily damage; the carbide steel blade is brittle and will shatter if it is dropped or abused; the spring steel tip in front of the saw blade is used for scraping grout out of corners where the saw blade cannot reach.

Grout Scrubbing Pad. A non-scratch nylon pad impregnated with abrasive used for cleaning grout from the surface of tile.

Growth Ring. The layer of wood growth put on a tree during a single growing season; annual growth ring.

Grubbing. The act or process of clearing and digging up roots and stumps.

GSU. Glazed Structural Unit.

GTA. Glass Tempering Association.

GTAW. Gas Tungsten-Arc Welding.

Guarantee. 1.Written or implied assurances for a specific part of the project, or for the project as a total. 2. An undertaking or document stating that a thing will or will not happen.

Guaranteed Maximum Price (GMP). An amount stipulated in a construction contract as the maximum sum payable by the owner to the contractor for the work specified.

Guarantor. One who gives a guarantee.

Guarded. Covered, shielded, fenced, enclosed, or otherwise protected by suitable means to remove the liability of dangerous contact or approach by persons or objects to a point of danger.

Guardrail. 1. A safety railing used as a barrier to prevent encroachment or accidental falling from heights. 2. In street or highway construction, a barrier to keep vehicles in their lanes. 3. A device for protecting a machine part or the operator of a machine.

Guardrail Removal. The act or process of removing a protective or safety device.

Guesstimate. An estimate based on a mixture of guesswork and calculation.

Guest. Any person hiring or occupying a room for living or sleeping purposes.

Guest Room. Any room or rooms occupied, or intended or designed to be occupied by a guest for sleeping purposes.

Guide Coat. In painting, a coat similar in composition to the finish or color coat, but of a different color to help obtain complete coverage.

Gum. A nonvolatile viscous plant exudate which either dissolves or swells up in contact with water; many substances referred to as gums, such as pine and spruce gum, are actually oleoresins; used in making varnishes and paints.

Gum Arabic. The dry gummy exudation of Acacia Senegal, a white powdered resin, is used in adhesives, cold water paint, and in show card colors.

Gum Turpentine. Oleoresinous material obtained from living pine trees; gum turpentine, when distilled, provides gum rosin and gum spirits of turpentine.

Gun Distance. Space between tip of spray gun and work.

Gunite. A concrete material applied by pumping through a hose; also called dry-mix Shotcrete.

Gusset Plate. A flat steel plate to which the chords of a truss are connected at a joint; a stiffener plate.

Gutta-Percha. A tough plastic substance obtained from the latex of various Malaysian trees, resembles rubber but contains more resin, is used as insulation and other uses.

Gutter. A channel to collect rainwater and snow melt at the eaves of a roof; a shallow channel constructed of steel, copper, aluminum, or plastic compounded with other materials to increase strength and wearing or rust resistance, positioned just below and along the eaves of a building for collecting and diverting water from a roof.

Gutter Nail. A long nail used for hanging rainwater gutters on light wood framed residences.

Gutter Removal. The act or process of tearing down and carrying away an old gutter system.

Gutter Strap. Metal band used to support the gutter.

Guy Cable. A wire used to secure a tall exterior mast, antenna, or other structure in place.

Guy Rod. A metal rod with a cable or rope attached, leading to an object to support and stabilize it.

Gym. Gymnasium.

Gym Curtain. A vertical piece of fabric used to divide a gymnasium into separate areas.

Gym Floor. A level surface, usually made of a resilient hardwood, used for events and sports that take place in a gymnasium.

Gym Locker. A lockable compartment used to store personal effects while a person is using gym facilities.

Gymnasium. A room or building used for indoor sports.

Gymnasium Floor Anchor Channel. Metal channel used to anchor wood gymnasium flooring.

Gymnasium Underfloor Insulation. Insulation, usually board insulation, located between sleepers below gymnasium wood flooring to dampen the sound drum effect.

Gym Scoreboard. A large board found in a gymnasium, which displays the score and often other information of a game or match.

Gypsite. An earthy deposit found at or near the surface of the ground, consisting of finely crystalline gypsum mixed with loam, clay, sand, and humus; gypsum content generally ranges from 60% to more than 90%.

Gypsum. A common naturally occurring mineral composed of hydrous calcium sulfate, $CaSO_4 2H2O$; the main component of sheetrock or drywall; gypsum, when heated, forms plaster-of-Paris.

Gypsum, Alpha. See Alpha Gypsum.

Gypsum Backing Board. A 1/4 (6.35 mm)to 5/8 inch (15.875 mm) thick gypsum board for use as a backing for gypsum wallboard, acoustical tile or other dry cladding.

Gypsum Block. A cast gypsum building unit; also called a Gypsum Tile.

Gypsumboard. An interior facing panel consisting of a gypsum core sandwiched between paper faces; also called Drywall, Plasterboard, or Sheetrock.

Gypsum Concrete. A calcium gypsum mixed with wood chips, or aggregate, or both, used primarily for poured roof decks.

Gypsum Core Board. A 3/4 (19.0 mm) to 1 inch (25.4 mm) gypsumboard consisting of a single board or factory laminated multiple boards used as a gypsum stud or core in semisolid or solid gypsum board partitions.

Gypsum Fill. Troweled on plaster material to make depressions level.

Gypsum Formboard. A gypsumboard used as the permanent form for poured gypsum roof deck.

Gypsum Gauging Plaster. A plaster for mixing with lime putty to control the setting time and initial strength of the finish coat; classified either as quickset or slowset.

Gypsum High Strength Basecoat Plaster. A gypsum cement for use with sand aggregate to achieve high compressive strength plaster.

Gypsum Lath. A plaster base manufactured in the form of sheets or slabs of various sizes and either 3/8 or 1/2 inch thick, having an incombustible core, essentially gypsum, and surfaced with special paper suitable for receiving gypsum plaster.

Gypsum Masonry. Molded, lightweight masonry units made from gypsum plaster, water and fiber.

Gypsum Molding Plaster. A specially formulated plaster used in casting and ornamental plasterwork; may be used neat or with lime.

Gypsum Neat Plaster. A plaster requiring the addition of aggregate on the job; it may be unfibered or fibered (vegetable, or glass fibers).

Gypsum Plaster. Plaster in which the cementing substance is gypsum.

Gypsum Ready Mixed Plaster. A plaster which is mixed at the mill with a mineral aggregate and may contain other ingredients to control time of set and working properties; only the addition and mixture of water is required on the job; also called Mill-Mixed and Pre-Mixed

Gypsum Roof Deck. A lightweight roofing substrate made of gypsum in the form of structural boards or poured in place over a structural deck.

Gypsum Sheathing. Flat sheet material of gypsum board with water repellent paper secured to exterior side of walls, roof or floor framing used to create rigidity in building superstructure, serve as base to receive other construction, and add fire resistive characteristics; not designed for long term direct exposure to the elements.

Gypsum Tile. See Gypsum Block.

Gypsum Trowel Finish. Various proprietary ready-mixed finish coat materials consisting essentially of calcined gypsum.

Gypsum Wall Sheathing, Fire-Rated. Flat sheet material of fire rated gypsum board.

Gypsum Wallboard. A gypsum board used primarily as interior surfacing material for building structures.

Gypsum Wood-Fibered Plaster. A mill-mixed plaster containing a small percentage of wood fiber as an aggregate, used for fireproofing and high strength.

Gyrate. To go in a circle or spiral; revolve, whirl.

H

Hack. 1. To cut with irregular or unskilled blows. 2. To cut back and roughen a plastered or other surface.

Hacking. Laying masonry units so that the bottom edge is set back from the plane surface of the wall.

Hacksaw. Small toothed handsaw for sawing metals.

Hairline Cracking. 1.Tiny cracks in a surface. 2. Small cracks of random pattern in an exposed concrete surface. 3. Very narrow cracks in a paint or varnish film.

Hairpin Bars. Bars, usually small sizes, bent to a hair pin shape and used for such purposes as short hooked spacer bars in columns and walls and for special dowels.

Hair Trap. A device to trap and retain the hair content of waste water and sewage, as in a barber shop or beauty salon.

Half and Half. A dry or dampened mixture of one part portland cement and one part extra-fine sand; used as a filler in the joints of mounted ceramic mosaic tiles to keep them evenly spaced during installation.

Half Bath. A room containing a toilet and a lavatory.

Half Block. A concrete masonry unit that is half the length of a standard unit.

Half Round Gutter. A narrow rounded channel from one piece of metal, similar to a pipe cut longitudinally or along its length, used as a gutter.

Half Round Molding. An ornamental strip having one flat side and one rounded side.

Half-Timber. A form of construction composed of timber framing with the spaces filled in with brick or plaster.

Halide Fixture. A lighting fixture with a Halide Light.

Halide Light. An electric-discharge lamp that makes its light from a metal vapor such as sodium or mercury.

Halide Refrigerants. Family of refrigerants containing halogen chemicals.

Halide Torch. Type of torch used to safely detect halogen refrigerant leaks in system.

Hall. 1. Lobby or entrance room. 2. A university building for a special purpose. 3. A hallway or corridor.

Hallway. A corridor or hall.

Halogens. Any of the group of poisonous non-metallic elements, fluorine, chlorine, bromine, iodine, and astatine, which form halides by simple union with a metal, as in sodium chloride.

Halon System. A system using halon gas for the fire protection of water sensitive equipment.

Hammer. A handtool consisting of a solid head set crosswise on a handle and used for pounding.

Hammered Effect Finish. So called because of its resemblance to hammered metal; produced by incorporating an aluminum powder in vehicle which controls leafing and nonleafing effect to create unique designs.

Hammer Fracture. A tear in the gypsum board face and core caused by improperly hitting or over driving the nail; also called Nail Fracture.

Hammer Loss. A test to determine the impact resistance of gypsum board.

Hand. The feel of a carpet in the hand; determined by such factors as pile height, quality and kind of fibers, type of construction, type of backing, and dimensional stability.

Hand Buggy. A hand pushed or pulled two or four wheeled cart, used to transport small amounts of concrete or material.

Hand Dryer. A mechanical/electrical device that blows hot air over wet hands to dry them.

Handed Locks. Locks that must be used on the same handed door, as opposed to reversible locks.

Hand Excavation. The act or process of digging out earth using hand tools.

Handhole. 1. Access hole used for repair and cleaning. 2. An enclosure installed in the earth, deck, floor of a building, or similar location and used as a pull or junction box for underground electrical or communication conductors; the enclosure is provided with a removable cover and so designed that the conductors may be pulled, spliced, or otherwise handled without requiring a person to enter the enclosure.

Handicapped Parking. A parking space designated as reserved for the vehicles of disabled persons.

Handicapped Partition. A dividing wall that forms a toilet cubicle area that is accessible to the disabled; commonly larger than a standard cubicle, large enough for wheelchair access.

Handicapped Plumbing. Plumbing devices and layouts specifically designed for use by disabled individuals.

Handicapped Symbol. The profile of a wheelchair and occupant placed on a sign to denote access for disabled persons.

Hand Level. A hand-held viewing tube with a spirit level attached to fix a level line of sight.

Handling, Air. See Air Handling System.

Handling, Rubbish. See Rubbish Handling.

Handling, Trash. See Trash Handling.

Handling, Waste. See Waste Handling.

Hand of a Lock or Door. Indicates the direction of swing or movement and/or locking security side of a door; designated as left hand, left hand reverse, right hand, or right hand reverse.

Hand Prints. Wallpapers printed by hand, usually with the silk-screen process.

Handrail. Member which is normally grasped by hand for support at stairways and other places where needed for safety of pedestrians.

Handrail Bracket. A support for a handrail.

Handrail, Pipe. See Pipe Handrail.

Handsaw. A wood saw which is powered by hand, commonly called a Carpenter's Saw or Wood Saw.

Hand Seeding. The act or process of planting seed by hand.

Handsplit Shingle. A shingle made by splitting a block of wood, usually cedar or redwood, along its grain and thereby creating a shingle which may be used for roofing or siding; also called Shakes or Shake Shingles.

Handsplit Shingles. See Cedar Shake.

Hang. To install certain materials such as gypsum lath, gypsum wallboard, or wallpaper.

Hangar. A shelter for storing and repairing aircraft.

Hanger. 1. A tradesman who applies gypsum board products. 2. Vertical-tension member supporting a load. 3. Device attached to walls or other structure for support of pipe lines.

Hanger, Door Rail. See Door Rail Hanger.

Hanger, Joist. See Joist Hanger.

Hanger Rod. A rod for connecting pipe, gutters, or ceiling framework to a support.

Hanger's Bench. See Bench, 4.

Hanger's Tee. See Tee Square.

Hanger Wire. Malleable wire for hanging ceilings and equipment from the structure above.

Hansa Yellow. A family of organic yellow pigments.

Hardboard. A very dense panel product, usually with at least one smooth face, made of highly compressed wood fibers; consolidated under heat and pressure, and having a density factor of approximately 50 to 80 pounds per cubic foot.

Hardboard Door. A door constructed of compressed wood fibers.

Hardboard Panel Siding. Panel siding made of hardboard.

Hardboard Panel. Flat sheet material of fibers consolidated under heat and pressure in hot press.

Hardboard Soffit. Hardboard finish material covering the underside of a part or member.

Hardboard Underlayment. Flat sheet material of hardboard placed over subflooring to provide smooth and even surface to receive finish floor covering.

Hard-Burned. Clay products which have been fired at high temperatures to near vitrification, generally producing relatively low absorption and high compressive strengths.

Hard Edge. A special core formulation used along the paperbound edges to improve resistance to damage during the handling and application of gypsum board.

Hardener. Curing agent; promoter; and catalyst used in making and finishing concrete.

Hardening. The gain of strength of plaster, mortar, or concrete after setting; see Set.

Hardhat. 1. A safety helmet of metal or plastic worn by workers on a construction site. 2. A construction worker who wears such a hat.

Hard Money. The actual cash paid to a seller in addition to any promissory notes.

Hardness. 1. A property of a material that resists indentation. 2. Of water, the quantity of dissolved minerals, such as calcium and magnesium compounds.

Hard Oil Finish. A varnish giving the effect of a rubbed-in oil finish but producing a hard surface; any interior architectural varnish with a moderate luster.

Hardpan. Hard, tight soil; a hard layer that may form just below plow depth on cultivated land.

Hardscape. The non-organic elements of a landscape development, such as walls, paving, fences, and lighting standards.

Hardstand. A hard surfaced area for parking an airplane.

Hard Tile. A term used in the tile trade to designate types of tile, such as ceramic, glass mosaic, or marble tile, over which the tile trade has jurisdiction, as compared to resilient tile.

Hardwall. 1. Base coat plaster. 2. Regionally the term differs; in some cases it refers to sanded plaster, while in others to neat.

Hardware. The miscellany of metal or plastic fittings and fastenings used in or on a building and its parts.

Hardware Cloth. Steel wire mesh, usually galvanized, commonly 1/8-, 1/4-, or 1/2-inch mesh.

Hard Water. Water with an excessive mineral content.

Hardwood Dimension Lumber. See Dimension Stock.

Hardwood Plywood Panel. Plywood with face plies manufactured from hardwood species such as cherry, oak, birch, ash, walnut, maple, gum, mahogany, or teak.

Hardwoods. Generally one of the botanical groups of trees that have broad leaves in contrast to the conifers or softwoods. The term has no reference to the actual hardness of the wood

Hardy Cross Method. See Moment Distribution.

Harmon Hinge. A hinge designed to swing a door into a pocket at a right angle with the frame.

Harsh Mortar. A masonry mortar that is hard to spread under the trowel, usually because of a lean ratio of cement paste to aggregates.

Hasp Assembly. A fastener for a door or lid consisting of a hinged metal strap that fits over a loop and is secured by a pin or padlock.

Hatch. A hinged or removable cover in a floor or roof which permits ventilation or the passage of persons or objects.

Hatchet. A multi-purpose short handled hand tool with a head suitable for alternative use as a hammer or axe; the convex head configuration to properly dimple the nail and the hatchet blade used to cleave and adjust framing and wallboard edges to fit, or as a jacking wedge.

Hatch, Roof. See Roof Hatch.

Hatchway. An opening giving access usually by ladder to a lower space like a cellar.

Haul Resistance. The distance measured along the most direct practical route between the center of the mass of excavation and the center of mass of the fill as finally placed; the distance material is moved.

Haul, Station Yards of. Station Yards of Haul.

Hawk. A metal square with a wooden handle at the center, used by a plasterer or tilesetter, to temporarily hold mortar or plaster; commonly, a rubber pad fits over the handle and covers that portion of the metal hawk that would come in contact with the hand.

Hazardous Waste. A material or substance characterized by a propensity to be unhealthy, toxic, or dangerous.

Hazardous Wiring. All wiring except that which conforms with all applicable laws in effect at the time of installation and which has been maintained in good condition and is being used in a safe manner.

HB. Hose Bibb.

H Beam. A steel beam which in cross section resembles the letter H; commonly used in earthwork as a retaining structure or piling.

Head. 1. The operative part of a tool. 2. The flattened part of a nail. 3. The upper horizontal part of a door, window, or other opening. 4. The pressure produced by a difference in elevation of two points in a body of fluid. 5. The top part of a column or pillar.

Header. 1. Length of pipe or vessel to which two or more pipe lines are joined carries fluid from a common source to various points of use; a manifold. 2. The pipe that runs across the top or bottom of an absorber plate, gathering or distributing the heat transfer fluid from or to the grid of pipes that runs across the absorber surface. 3. A lintel. 4. A structural support over an opening. 5. A joist that supports other joists; a short joist into which the common joists are framed around or over an opening. 6. A brick that is laid on its flat surface across a wall and will show only its end on the surface or face of the wall. 7. A masonry unit which overlaps two or more adjacent wythes of masonry to tie them together; also called a Bonder.

Header Bond. A brick bond that will show only headers on the face or surface of a wall; each header would be divided evenly over the header underneath it.

Header, Ceiling. A joist perpendicular to and supporting ceiling joists.

Header Course. A brick bond in which a course made up of headers only.

Header High. The height up to the top of the brick course that is directly under a header course.

Head Flashing. The exterior flashing installed at the head of a door, window, or other opening.

Head Friction. Head required to overcome friction of the interior surface of a conductor and between fluid particles in motion.

Head Joint. See Butt Joint, 4.

Headlap. The minimum distance, measured at 90 degrees to the eave along the face of a shingle or felt as applied to a roof, from the upper edge of the shingle or felt, to the nearest exposed surface.

Head, Pop-Up. See Pop-Up Head.

Head Pressure Control. Pressure-operated control which opens electrical circuit if high-side pressure becomes too high.

Head Pressure. Pressure in the condensing side of a refrigerating system.

Headroom. The clear space between the floor line and ceiling, as in a stairway.

Head, Service. See Service Head.

Head, Static. See Static Head.

Head, Total Static. See Total Static Head.

Head Velocity. Height of fluid equivalent to its velocity pressure in a flowing fluid.

Headwall. A wall, usually constructed of concrete or masonry, that is placed at the inlet side of a drain or culvert to protect fill from scouring, undermining, or to divert flow.

Headwall, Removal. The act or process of demolition and removal of a headwall.

Healing Power. The ability of a glaze to heal surface blemishes during firing.

Hearing. A proceeding conducted by a judge or arbitrator who receives evidence about a dispute.

Hearsay. Evidence of an event that the witness did not personally perceive.

Hearth. The masonry floor of a fireplace and the floor surface in front of a fireplace.

Heart Rot. Any rot characteristically confined to the heartwood; it generally originates in the living tree.

Heartwood. The wood extending from the pith to the sapwood, the cells of which no longer participate in the life processes of the tree; heartwood may contain phenolic compounds, gums, resins, and other materials that usually make it darker and more decay resistant than sapwood.

Heat. 1. The condition of being hot. 2. A form of energy arising from the random motion of the molecules of bodies, which may be transferred by conduction, convection, or radiation.

Heat-Affected Zone. That portion of the base metal which has been altered by the heat of welding, cutting, or brazing.

Heater. Any heat-producing appliance or unit, such as a furnace.

Heater Duct. The conduit in a heating system that conveys warmed air throughout a building or structure.

Heater, Electric. See Electric Heater.

Heater, Unit. See Unit Heater.

Heater, Water. See Water Heater.

Heat Exchanger. A device for the transfer of heat from one medium to another.

Heating Coil. An element where heat is transferred to an air stream; the heating medium may be steam, hot water, or electrical.

Heating Control. Device which controls the temperature of a heat transfer unit.

Heating Value. Amount of heat which may be obtained by burning a fuel; usually expressed in BTUs per pound, BTUs per gallon, or calories per gram.

Heat Intensity. Heat concentration in a substance as indicated by the temperature of the substance through use of a thermometer.

Heat Lag. The time it takes for heat to travel through a substance heated on one side.

Heat Leakage. Flow of heat through a substance.

Heat Load. Amount of heat, measured in BTUs or watts, which is removed during a period of 24 hours.

Heat Loss. The heat lost from a building through conduction, convection, and radiation as well as through air infiltration and ventilation systems.

Heat of Compression. Mechanical energy of pressure changed into energy of heat.

Heat of Fusion. Heat released from a substance to change it from a liquid state to a solid state; the heat of fusion of ice is 144 BTUs per pound (151.9 joules).

Heat of Hydration. The thermal energy given off by concrete, masonry, or gypsum as it cures.

Heat of Repression. 1. Process by which oxygen and carbohydrates are assimilated by a substance. 2. When carbon dioxide and water are given off by a substance.

Heat Pipe. A method of heat recovery from exhaust air in which a coil has a partition wall dividing the coil face into two sections; this system has no moving parts and relies on the slope of the coil to operate, sloping down from the cold end to the hot end.

Heat Pump. A combination heating and cooling device. In the winter it extracts heat from air as cold as 20 degrees, and in the summer it works in reverse to become an air conditioner.

Heat Recovery. The extraction of heat from a source not primarily designed to produce heat.

Heat Sensor. A battery operated or hard wired mechanical/ electrical device that either makes a loud noise or flashes a light when raised air temperature is detected; for use as a substitute for a smoke detector in locations that might have ambient smoke such as in kitchens.

Heat-Set Nylon. Nylon fiber which has been heat treated to retain a desired shape.

Heat Sink. A substance or device for the absorption and dissipation or recovery of heat in connection with a passive solar heating system.

Heat/Smoke Vent. A vent or chimney forming a passageway for expulsion of vent gases from gas-burning units to the outside air.

Heat Storage. A device.or medium that absorbs collected solar heat and stores it for periods of cloudy or cold weather.

Heat Strengthened Glass. Glass that has been strengthened by heat treatment, though not to as great an extent as tempered glass.

Heat Time. In multiple-impulse welding or seam welding the time during which the current flows during any one impulse.

Heat Transfer. Movement of heat from one body or substance to another; heat may be transferred by radiation, conduction, convection, or a combination of these three methods.

Heat Wheel. A method of heat recovery from exhaust air in which both the intake and exhaust ducts are connected to the wheel.

Heavy Bodied Oil. A high viscosity oil.

Heavy Centered Pattern. Spray pattern having most paint in center, less at edges.

Heavy Duty Tile. Tile suitable for areas where heavy pedestrian traffic is prevalent.

Heavy Soil. A fine grained soil made up largely of clay or silt.

Heavy Timber. Construction requiring noncombustible exterior walls with a minimal fire-resistance rating of two-hours, laminated or solid interior members, heavy plank, or laminated wood floors and roofs.

Hectare. A unit of land area equaling 10,000 square meters or 2.471 acres.

Heel. 1. The point at the end of a truss where the top and bottom chords intersect. 2. See Steel Square.

Heeling In. Temporary planting of trees and shrubs.

Heel, Rafter. The end or foot that rests on the wall plate.

Hegeman Grind Gauge. Proprietary instrument for measuring smoothness of pigment dispersion in liquid paint.

Height. The dimension measured at right angles to the direction of the thickness and length of a masonry unit, the vertical dimension, as used in a wall.

Helmet. A protective device used in arc welding for shielding the face and neck; a helmet is equipped with a suitable filter glass and is designed to be worn on the head.

Hemihydrate. A hydrate containing half a molecule of water to one of the material forming the hydrate; used to describe the form of calcined gypsum generally used for plaster.

HEPA. High Efficiency Particulate Arrestance; see HEPA Filter.

HEPA Filter. High-Efficiency Particulate Air Filter; such filters are rated to trap at least 99.9% of all particles 0.3 microns in diameter or larger.

Hermetic Compressor. Compressor which has the driving motor sealed inside the compressor housing; the motor operates in an atmosphere of the refrigerant.

Hermetic Motor. Compressor drive motor sealed within same casing which contains the compressor.

Hermetic Seal. An airtight seal.

Hermetic System. Refrigeration system which has a compressor driven by a motor contained in the compressor dome or housing.

Heroic Scale. A statue of colossal size or proportions.

Herringbone Blocking. Wood blocking between studs set at a slight angle in alternate directions in adjacent stud spaces, the angle making it possible to drive the blocks in tightly between studs.

Herringbone Bond. Bricks laid in a zigzag fashion representing a herringbone pattern.

Herringbone Floor. A floor covering made up of rows of parallel lines with adjacent rows slanting in reverse directions.

Hertz (Hz). A measurement of the frequency of vibration of sound measured in cycles per second; one cycle per second equals one Hertz.

Hessian. See Burlap.

Heteropolymer. The result of Heteropolymerization.

Heteropolymerization. See Polymerization.

Heuristic. By trial and error and experiment rather than by science.

Hew. To fell, chop, or shape with an axe

Hex. 1. Hexagon. 2. Hexagonal.

Hexagon. A polygon of six sides.

Hex Nut. A six-sided, short metal nut with a threaded hole for receiving a rod or threaded bolt.

Hg. Mercury.

HI. Hydraulic Institute.

Hickey. A tool for bending pipe, conduit, and reinforcing bars.

Hiding Bond. Ability of a paint to hide or obscure the surface on which it is applied; degree of opacity of a pigment or paint.

Highbay Lighting. A lighting system located high above work or floor level.

High Boiling Solvent. A solvent with an initial boiling point above 302° F. (150° C.) such as diacetone alcohol or cellosolve acetate.

High Build. Producing thick dry films per coat.

High Calcium Lime. A type of lime containing principally calcium oxide or hydroxide and not more than 5% magnesium oxide or hydroxide.

High Chair Reinforcing. A chair-shaped device used to hold steel reinforcement off of the bottom form of a slab while the concrete is being poured.

High Early Strength Cement. Cement that develops strength more quickly than ordinary cement, typically will be as strong at 3 days and 7 days as normal cement would in 7 days and 28 days; alumina cement.

Highest and Best Use. The use of land that will produce the highest future net income.

High Flash Naphtha. Aromatic solvent having a high flash point, minimum 113° F, 45° C.

High Frequency Curing. See Radio Frequency Curing.

High Joint. See Crown, 1.

High Lift Grouting. A method of construction in which concrete block units may be laid the entire height of the wall before grouting; this method of construction requires special inspection.

Highlight. A smooth area, noticeably different from the normal surface.

Highlighting. Making certain parts of finished project appear lighter than other parts.

High Magnesium Lime. A type of lime containing more than 5% magnesium oxide or hydroxide.

High-Pressure Boiler. Boiler furnishing steam at pressures of 15 pounds per square inch gauge or higher (1.05 kg/cm^2).

High-Pressure Cut-Off. Electrical control switch operated by the high-side pressure which automatically opens electrical circuit if too high a pressure is reached.

High Pressure Gauge. Instrument for measuring pressures in range of 0 psi to 500 psi (0 kg/cm^{2}to 3.52 kg/cm^2).

High Pressure Steam Curing. See Autoclave Curing.

High Rate Sand Filter. A sand filter designed for flows in excess of five gallons per minute per square foot, but not in excess of 20 GPM per square foot.

High Relief. Sculptural relief in which at least half of the circumference of the modeled form projects.

High-Rise. A multi-storied building with elevators.

High Side. Parts of a refrigerating system which are under condensing or high-side pressure.

High-Side Float. Refrigerant control mechanism which controls the level of the liquid refrigerant in the high-pressure side of mechanism.

High Strength Bolt. A bolt designed to connect steel members by clamping them together with sufficient force that the load is transferred between them by friction.

High Tension. High Voltage.

High-Vacuum Pump. Mechanism which can create a vacuum in the 1000 to 1 micron range.

High Voltage Cable. Cable manufactured to withstand high-voltage.

High Voltage Circuit. Any circuit having a difference of potential of more than six hundred volts, or seven hundred fifty volts where specified in certain regulations, between any two conductors of the circuit.

High Voltage. Electrical power of over 600 volts.

High Wall. Undisturbed soil or rock bordering a cut; a face which is being excavated, as distinguished from spoil piles.

Highway Demolition. The destruction and removal of a public road.

Hinge. 1. A joint fixing the relative position of the ends of two or more structural members, but permitting their relative rotation. 2. A piece of door hardware that permits the opening and closing of a door by joining the door to the jamb with a flexible device.

Hinge Stile. The door stile to which the hinges are installed as distinguished from the lock stile.

Hip. The diagonal intersection of planes in a hip roof.

Hippodamian Plan. Greek plan, derived its name from Hippodamos, with grid city blocks, informal organization of public spaces and facilities within the grid matrix; better known examples are the Ionian cities of Ephesus and Miletus, 5th Century BC.

Hip Rafter. A structural member of a roof forming the junction of an external roof angle or, where the planes of a hip roof meet.

Hip Roof. A roof consisting of four sloping planes that intersect to form a pyramidal or elongated pyramid shape; a roof which slopes up toward the center from all sides, necessitating a hip rafter at each corner.

Hod. A portable trough for carrying plaster, mortar, and bricks, fixed crosswise on top of a pole and carried on the shoulder.

Hoddability. A term descriptive of the ease with which a plaster mortar may be handled with a hod or hawk, dependent upon flow characteristics and angle of repose of the mortar.

Hoe. A long handled tool with a thin flat metal blade for cultivating, weeding, and mixing.

Hog Ring. A heavy galvanized wire staple applied with a pneumatic gun which clinches it in the form of a closed ring around stud, rod, pencil rod or channel.

Hoist, Industrial. See Industrial Hoist.

Hoistway. Any shaftway, hatchway, well hole, or other vertical opening or space in which an elevator or dumbwaiter is designed to operate.

Holder, Door. See Door Holder.

Hold Harmless. See Indemnification.

Holding Period. In the manufacture of concrete products, the period between completion of casting and the introduction of additional heat or the steam curing period.

Hold Out. Ability to prevent soaking into substrate.

Holdover Tenant. A tenant that does not vacate the property at the end of a lease term.

Hole. 1. An opening or perforation in a material. 2. A pit in the ground.

Holiday. 1. In roofing, a skipped area of liquid applied roofing. 2. In painting, unintentional missing of an area of surface being painted.

Hollow-Back. Removal of a portion of the wood on the unexposed face of a wood trim member to more properly fit any irregularity in bearing surface.

Hollow Block. Concrete blocks that can be filled with insulation or reinforced and grouted.

Hollow Clay Tile. A tile building material manufactured in a variety of sizes and forms, used for both exterior walls and partitions.

Hollow Column. A vertical column constructed so as to create an air space within.

Hollow Concrete Masonry. Concrete masonry units that are manufactured with open cores, such as ordinary concrete blocks.

Hollow-Core Construction. A panel construction with faces of plywood, hardboard, or similar material bonded to a framed-core assembly of wood lattice, paperboard rings, or the like, which support the facing at spaced intervals.

Hollow Core Door. A flush hardwood or plywood door glued to a skeletal framework with the interior remaining void or honeycombed.

Hollow Core Slab. A precast concrete slab element that has internal longitudinal cavities to reduce its self-weight.

Hollow Masonry Unit. A masonry unit whose net cross-sectional area in any plane parallel to the bearing surface is less than 75% of its cross-sectional area measured in the same plane.

Hollow Metal Door. A hollow core door constructed of sheet metal that has been channel-reinforced; the core is often filled with some type of lightweight material.

Hollow Metal Frame. A sheet metal door frame with reinforcing at the hinges and strikes.

Hollow-Tube Gasket. Sealing device made of rubber or plastic with tubular cross-section.

Hollow Wall. A wall built of solid masonry units laid in and so constructed as to provide an air space within the wall.

Homestead. 1. A tract of land granted by the U.S. upon recording, improving, and cultivating the property. 2. A residence. 3. A home and adjoining land occupied by a family; this can be exempt from judgment up to a certain amount by filing a Declaration of Homestead.

Hone. A honing stone.

Honeycomb. Voids in concrete resulting from segregation and poor consolidation at the time concrete is placed.

Honeycomb Core. A sandwich core material, for doors or building panels, constructed of thin sheet materials or ribbons formed to honeycomblike configurations.

Honeycombing. Checks, often not visible at the surface, that occur in the interior of a piece of wood, usually along the wood rays.

Honing Stone. A fine-grit stone for sharpening cutting tools.

Hood, Exhaust. See Exhaust Hood.

Hood, Range. See Range Hood.

Hook. A semicircular bend in the end of a reinforcing bar to develop bond, and thus anchorage.

Hooke's Law. Stress and strain are proportional up to the elastic limit of the material.

Hook Knife. A special curved knife used primarily for cutting resilient floor coverings; also called a Linoleum Knife.

Hook Stick. An implement that is curved or bent for holding, catching, or pulling.

Hook Strip. Strip of wood, plastic, or metal secured to a wall surface to provide a mounting surface for hooks, such as coat hooks or key hooks.

Hook Wire. A hook implement that is curved or bent for holding, catching, or pulling.

Hoop. Circular shaped pieces of steel reinforcing bar.

Hopped-Up Mud. Mortar mixed with an accelerator.

Hopper. Deep water basin with all four sides sloping for diving.

Hopper Window. A window whose sash pivots on an axis along the sill, and that opens by tilting toward the interior of the building.

Horizon. The line at which the earth and sky appear to meet.

Horizontal. Parallel to the horizon.

Horizontal Angle. An angle in a horizontal plane.

Horizontal Application. The application of gypsum board with the long dimension at right angles to the framing members; also referred to as perpendicular.

Horizontal Branch. A branch drain in the DWV system that extends from a soil, waste stack, or building drain to the fixture trap; it may or may not have vertical sections or branches off of it which receive discharge from one or more fixture drains and carry this discharge to the soil, waste stack, or house drain; also called a Lateral Branch.

Horizontal Broken Joints. A style of laying tile with each course offset one-half its length.

Horizontal Fixed Position. The position of a pipe joint wherein the axis of the pipe is approximately horizontal and the pipe is not rotated during welding.

Horizontal Forces. Any lateral loads imposed on a structure, including those caused by wind or earthquake forces.

Horizontal Hardboard Siding. Horizontal siding made of hardboard often used as clapboards on a house.

Horizontally Laminated Wood. Laminated wood in which the laminations are so arranged that the wider dimension of each lamination is approximately perpendicular to the direction of loa ᵈ.

Horizontal Rolled Position. In pipe welding, the position of a pipe joint wherein welding is performed in the flat position by rotating the pipe.

Horizontals. Reinforcing bars running horizontally.

Horizontal Shaft Mixer. A mixer having a stationary cylindrical mixing compartment, with the axis of the cylinder horizontal, and one or more rotating horizontal shafts to which mixing blades or paddles are attached.

Horizontal Siding. Linear horizontal material, usually overlapping, used as an exterior surface or cladding for an exterior framed wall.

Horizontal Wood Siding. Linear horizontal wood material, usually overlapping, used as an exterior surface or cladding for an exterior framed wall.

Horn. The extension of a stile, jamb or sill.

Horsepower. 1. A unit of electrical power in the U.S. equal to 746 watts. 2. In the English gravitational system, 550 foot-pounds of work per second.

Hose Bibb. Bibbcock. a water spigot or faucet with its nozzle threaded or coupling attached to accept a hose; also spelled Hose Bib; also called a Wall Hydrant or a Sill Cock.

Hose Reel. A round device upon which a hose is wound for easy dispensing.

Hospital Bed. A piece of furniture found in a hospital where patients recline, lie and sleep.

Hospital Cubicle. A small partitioned space or sleeping compartment in a hospital.

Hospital Divider. A movable partition in a hospital to create privacy.

Hospital Equipment. Apparatus or implements used in hospital applications.

Hospital Faucet. A water faucet that is controlled by feet to help eliminate the spread of germs by hands.

Hospital Panel. A partition in a hospital which divides areas into smaller spaces.

Hospital Stretcher. A narrow, movable bed with collapsible legs and wheels used for transporting patients, often from an ambulance to a hospital.

Hospital Wardrobe. A cabinet installed in a hospital patient room to store the patient's clothes; usually matches any other cabinet or furniture in the room.

Hot Aggregate Storage Bins. Bins that store the heated and separated aggregates prior to their final proportioning into the mixer.

Hotel. A building containing an establishment providing lodging and meals.

Hot Gas Bypass. Piping system in refrigerating unit which moves hot refrigerant gas from condenser into low-pressure side.

Hot Gas Defrost. Defrosting system in which hot refrigerant gas from the high side is directed through evaporator for short period of time and at predetermined intervals in order to remove frost from evaporator.

Hot Junction. That part of thermoelectric circuit which releases heat.

Hot Lacquer Process. Process where heat is used instead of volatile thinner to reduce the consistency of lacquer; hot lacquer can be applied with a higher percentage of solids than room-temperature lacquer.

Hot-Laid Plant Mixture. Plant mixes that must be spread and compacted while at an elevated temperature; to dry the aggregate and obtain sufficient fluidity of the asphalt (usually asphalt cement), both must be heated prior to mixing, hence the term hot-mix.

Hot Mud. Mortar mixed with an accelerator; also called Hot Stuff.

Hot Plate, Laboratory. See Laboratory Hot Plate.

Hot Spray. Spraying material heated to reduce viscosity.

Hot Stuff. 1. Mortar mixed with an accelerator; also called Hot Mud. 2. Hot bitumen.

Hot Tub. Refers specifically to a hydrotherapy unit normally constructed of wood designed and assembled in the traditional manner of tubs or casks, with side and bottoms formed of separate boards and the whole shaped to join together by pressure of the surrounding hoops, bands or rods as distinct from spa units formed of plastic, concrete, metal or other materials.

Hot-Water Boiler. A boiler furnishing hot water at pressures not more than 30 pounds per square inch gauge (2.12 kg/cm^2) or steam at pressures not more than 15 pounds per square inch gauge (1.06 kg/cm^2).

Hot Water Circulating Loop System. A piping arrangement in which hot water is constantly circulated in a loop from which individual plumbing fixtures are fed.

Hot Water Dispenser. A device used to heat and deliver small quantities of water.

Hot Water Heating System. System in which water is circulated through heating coils.

Hot Water Storage Tank. A tank that receives heated water and holds it for future distribution.

Hot Wire Relay. See Thermal Relay.

Hot Wire. 1. The live wire, carrying electric current, as opposed to the neutral or ground. 2. Resistance wire in an electrical relay which expands when heated and contracts when cooled.

House. A building for human habitation.

Housed Joint. A joint in which a piece is grooved to receive the piece which is to form the other part of the joint.

House Drain. The horizontal part of the drainage system which connects the DWV piping within the structure to the sanitary sewer or private sewage treatment equipment.

House Drainage System. Complete system of piping used for carrying away waste water and sewage; also called Building Drainage System.

House Paint, Outside. Paint designed for use on the exterior of buildings, fences and other surfaces exposed to the weather.

House Sewer. The piping that takes the soil and waste water from the building drain and conveys it to the public sewer or private sewage disposal system; also called Building Sewer.

House Trap. Located at the point at which the house drain leaves the building, designed to hold a quantity of water that prevents gasses from the sewer system from entering the building.

Housing. A groove cut at any angle with the grain and partway across the piece; used for framing stair risers and treads.

Howe Truss. A standard peaked roof truss configuration where the vertical web members and the bottom chord are in tension while the sloping web members and the top chord are in compression.

HP. High Pressure.

HPMA. Hardwood Plywood Manufacturer's Association.

HP Sodium Lamp. A high-pressure sodium vapor lamp that produces a wide-spectrum yellow light.

HPVA. Hardwood Plywood & Veneer Association.

Hub. 1. A surveyor's bench mark; used as a starting point for measuring elevations and horizontal distances. 2. The enlarged end of a hub and spigot cast-iron pipe.

Hub. See Bell.

Hub Union. A pipe fitting used to join two pipes without turning either pipe.

Hue. Color.

Human Relations. A behavioral aspect of management — handling of personnel and the consideration of each problem from the employee's point of view; the success or failure of the business often depends on the relationship between management and personnel.

Humidification. The addition of moisture to air, thereby increasing the latent heat.

Humidified Bond. In gypsum wallboard, the ability of the surfacing paper to resist delamination from the core under extremely high humidity conditions.

Humidifier. A device for supplying or maintaining humidity in a conditioned space.

Humidistat. Electrical control which is operated by changing humidity.

Humidity. The amount of water vapor in a given volume of air.

Humidity Ratio. The weight of water vapor per unit weight of dry air; also called Specific Humidity.

Hump Joint. See Crowned Joint.

Humus. Decayed organic matter; a dark fluffy swamp soil composed chiefly of decayed vegetation; also called peat.

Hung Slab. A nonstructural concrete slab having no bearing on walls, being totally supported from above by an overhead structural element.

Hurricane. A tropical cyclone with winds over 74 miles per hour; usually accompanied by rain, thunder, and lightning.

Hut. A small or temporary dwelling of simple construction.

Hutment. A collection of huts.

HVAC. Heating, Ventilation, and Air-Conditioning system.

HVAC Control. An apparatus that controls a heating, ventilation, and air conditioning system.

HW. Hot Water.

Hydrant. 1. A discharge pipe with a valve and threaded outlet from which water may be drawn from a water main; a faucet. 2. Fire hydrant; fire plug.

Hydrant Removal. The act or process of removing an old hydrant.

Hydrant, Wall. See Wall Hydrant.

Hydrate. 1. A chemical combination of water with another compound or an element. 2. Hydrated Lime.

Hydrated Lime. Calcium hydroxide, a dry powder obtained by treating quicklime with water; also called Hydrate.

Hydration. The process of adding water to a substance.

Hydraulic. Of a mechanism operated, moved, or effected by use of water pressure.

Hydraulic Cement. A type of portland cement that will harden or set under water.

Hydraulic Elevator. A lifting device powered by pressured fluid.

Hydraulic Excavator. A digging machine powered by pressured fluid.

Hydraulic Gradient. The slope of the surface of open or underground water.

Hydraulic Lift. An elevating device powered by pressured fluid.

Hydraulics. 1. The science of the conveyance of liquids through pipes. 2. The branch of physics having to do with the mechanical properties of water and other liquids in motion.

Hydraulic Spraying. (See Airless) Spraying by hydraulic pressure.

Hydrocarbon. An organic compound containing only carbon and hydrogen; fuels such as natural gas, gasoline, and coal contain hydrocarbons.

Hydrocarbon Resins. Obtained by catalytic polymerization of petroleum fractions.

Hydrochloric Acid. A solution of the colorless gas hydrogen chloride that is a strong corrosive irritating acid, used in dilute form for cleaning portland cement paste from brick or stone work.

Hydrofluoric Acid. An aqueous solution of hydrogen fluoride that is a weak poisonous acid that attacks silica and silicates, used for finishing and etching glass.

Hydrojet. A fitting which blends air and water creating a high velocity, turbulent stream of air enriched water.

Hydrojet Booster Pump System. A system whereby one or more hydrojets are activated by the use of a pump which is completely independent of the filtration and heating system of the spa.

Hydrology. The science of the properties of the earth's water, on the surface, subsurface, and atmospheric, and its movement in relation to land.

Hydrometer. A device for measuring the density of liquids.

Hydronic. Heating system which circulates a heated fluid, usually water, through baseboard coils by means of a circulating pump which is controlled by a thermostat.

Hydronic Heating System. A system that circulates heated water through convectors to heat a building or space.

Hydronics. Systems of heating or cooling that involve transfer of heat by a circulating fluid, as water or vapor, in a closed system of pipes.

Hydrophilic. Having an affinity for water; capable of uniting with or dissolving in water.

Hydrophobic. Having antagonism for water; not capable of uniting or mixing with water.

Hydroponics. The process of growing plants in sand, gravel, or liquid, without soil and with added nutrients.

Hydrostatic Pressure. Pressure exerted by standing water.

Hydrostatic Relief Valve. A valve installed in the main drain of the swimming pool, sometimes in other areas as well; when the pool is full, the valve is closed; when pool is empty, it is opened to allow ground water to flow into the pool; this prevents the pool from floating, which happens if ground water pressure builds up.

Hydrous. Containing water.

Hygrometer. An instrument, consisting of a dry bulb thermometer and a wet bulb thermometer, for measuring the humidity of the atmosphere or of a gas; also called a Wet and Dry Bulb Hygrometer or a Psychrometer.

Hygroscopic. 1. Tendency to absorb water. 2. Ability of a substance to absorb and release moisture and change physical dimensions as its moisture content changes. 3. A synthetic rubber (chlorosulfonated polyethylene) often used with neoprene in elastomeric roof coverings; Hypalon is a registered trademarked name of E.I. du Pont Co.

Hyperbolic Paraboloid Shell. A concrete roof structure with a saddle shape.

Hyperbolic Paraboloid. Double curved surface which can be generated by sliding a straight line along two other lines which are skew to each other but lie in parallel planes.

Hypocenter. The point below the epicenter at which an earthquake actually begins; also called the Focus.

Hypotenuse. The side of a right triangle which is directly opposite the right angle itself.

Hypothecate. To pledge property for repayment of a debt without giving up title or possession of the property

I

I. 1. Represents the moment of inertia of a member. 2. Occupancy Importance Factor. 3. Represents Amperes.

IAQ. Indoor Air Quality.

IBC. International Building Code.

I Beam. An American Standard designation for a particular section of hot-rolled steel which in cross section is shaped like a capital "I".

ICBO. International Conference of Building Officials.

ICC. International Code Council.

Ice Cream Cabinet. Commercial refrigerator which operates at approximately 0° F. (-18° C.); used for storage of ice cream.

Ice Dam. An ice obstruction along the eaves of a roof caused by the refreezing of water emanating from melting snow on the roof surface above.

Ice Maker. An appliance that freezes water into ice chunks or cubes.

Ice Melting Equivalent. The amount of heat absorbed by melting ice at 32° F. (0° C.) is 144 BTUs per pound of ice or 288,000 BTUs per ton; also called Ice Melting Effect

Ice Storage Bin. A container which holds ice and maintains its solid properties.

ID. 1. Inside Diameter. 2. Inside Dimension.

Idiot Stick. See Layout Stick. See Story Pole, 2.

Idler. Pulley used on some belt drives to provide proper belt tension and to eliminate belt vibration.

Ignition Transformer. Transformer designed to provide a high-voltage current; used in many heating systems to ignite fuel; it provides a spark gap.

IIAR. International Institute of Ammonia Refrigeration.

IIC. Impact Insulation Class.

IL. Intensity Level.

ILIA. Indiana Limestone Institute of America, Inc.

Illegal. Contrary to or against the law.

Illiquid. Deficient in liquid assets.

Illuminate. Light up, make bright, usually with electrical lighting.

Ilmenite. A mineral having the theoretical composition FO TiO_2used principally in the production of titanium oxide.

IMC. Intermediate Metal Conduit.

IME. 1. Ice Melting Equivalent. 2. Ice Melting Effect.

IMI. International Masonry Institute.

Immeasurable. Not measurable; immense.

Immiscibility. Incapability of being mixed or made homogeneous.

Immune. Not subject to legal process.

Impact. The effect that escalation or delay has on other work in a construction schedule, ripple effect.

Impact Barrier. A barrier constructed to resist dynamic loading on a surface.

Impact Insulation Class (IIC). A single figure rating that provides an estimate of the sound insulating performance of a floor or ceiling assembly.

Impasto. Thick application of pigment to canvas or other surface which makes the painting stand out in bold relief.

Impedance. The opposition to electrical flow in an alternating current electrical circuit.

Impeller. A blade of a pumping device; a propeller type fitting which produces suction when rotated.

Imperial Gallon. Unit of volume measure used in Great Britain and Canada, containing 277.42 cubic inches and weighing 10 pounds; compared with U.S. gallon of 231 cubic inches and 8.33 pounds in weight.

Impermeability. The resistance an asphalt pavement has to the passage of air and water into or through the pavement.

Impermeable. Does not permit the passage of fluids; impervious.

Impervious. Not letting water or moisture pass through or be absorbed; the degree of vitrification evidenced visually by complete resistance to dye penetration; impervious generally signifies zero absorption, except for floor and wall tile which are considered impervious up to 0.5 percent water absorption.

Imperviousness. The quality of resisting moisture penetration.

Impervious Tile. Tile designated by water absorption of 0.5 percent or less.

Implied Contract. A contract that is not in words.

Implied Covenant. A promise that is not expressed in a contract, but that is implied from the surrounding circumstances.

Implied Indemnity. An obligation to indemnify that arises not from the words of a contract, but from the circumstances of the parties.

Implied Warranty. A promise, usually related to a the quality or serviceability of goods, that is not in words but is implied from the circumstances of a sale.

Imported Marble. Marble shipped in from another country or area.

Impossibility. A doctrine of contract law that excuses performance that becomes physically impossible.

Impreg. Wood in which the cell walls have been impregnated with synthetic resin so as to reduce greatly its swelling and shrinking; impreg is not compressed; compare with Compreg.

Impregnate. To saturate, fill, or permeate one substance with another, as wood with chemicals.

Imprest Cash Fund. A petty cash fund maintained for minor disbursements and from time-to-time restored to its original amount by a transfer from the general cash account of an amount equal to the sum of disbursements.

Improvements. Any enhancements to value of property including buildings, paving, fencing, and landscaping

In Situ. 1. In the natural or original place. 2. Concrete poured in place.

In Situ Concrete. Poured in place concrete.

Inactive Door. In a pair of doors, the leaf that does not contain a lock, but is bolted when closed, and to which the strike is fastened to receive the latch or bolt of the active door.

Incandescent Fixture. An electric light fixture in which the lamp or bulb has a filament that gives off light when heated to incandescence by an electric current.

Incandescent Lamp. A device that produces light through an electrically heated filament.

Inch. A unit of linear measure equal to 1/12 of a foot and equal to 2.54 centimeters.

Inchoate. Only partly formed, or just begun, as an idea or the beginnings of an operation.

Incinerator. A furnace or apparatus for reducing waste materials to ashes by burning.

Incinerator, Laboratory. See Laboratory Incinerator.

Incinerator, Medical. See Medical Incinerator.

Incised. 1. Cut into the surface; engraved. 2. In ceramics, decorated by cutting or indenting the ware surface.

Incline. The slope of a roof expressed either in percent or in the number of vertical units of rise per horizontal unit of run.

Incombustible. Incapable of being burned.

Income & Expense Statement. Summarizes the operations of the company over a period of time; also known as a profit and loss statement.

Income Approach. A real estate appraisal method by which the income stream is capitalized.

Incompatibility. The quality or state of two materials that are not suitable to be used together.

Incompatible. Describes material which cannot be mixed with another material without damaging original properties.

Incomplete Fusion. Fusion which is less than complete.

Increaser. A fitting in the vent stack before the stack enters the roof; installed to enlarge the stack and reduces the possibility of water vapor condensing and freezing to the point of closing the opening.

Increaser. A pipe fitting which increases from a smaller pipe to a larger pipe.

Increment Borer. 1. An augerlike instrument with a hollow bit and an extractor, used to extract thin radial cylinders of wood from trees to determine age and growth rate. 2. Also used in wood preservation to determine the depth of penetration of a preservative.

Incubator. An apparatus for the maintenance of controlled conditions.

Incubator, Laboratory. See Laboratory Incubator.

Incurable Depreciation. Decreases to value of property that cannot be rectified or are too expensive to allay.

Indemnification. A contractual obligation by which a person or entity agrees to reimburse others for loss or damage arising from specified liabilities; also called Hold Harmless.

Indemnity. A promise to hold a person harmless from liability or loss.

Indentation. In a spot, seam, or projection weld, the depression on the exterior surface or surfaces of the base metal.

Independent Contractor. A person who, in performing services for another, is responsible only for the final result, and is not subject to control as to the methods used to achieve that result.

Indeterminate. Statically indeterminate

Indian Red. Red pigment made artificially by calcining copperas; has excellent permanency, is non-bleeding, alkali- and acid-fast.

Indicator Button. A device on a hotel lock to indicate whether or not the room is occupied.

Indicator Compounds. Chemical compounds, such as carbon dioxide, whose presence at certain concentrations may be used to estimate certain building conditions, for example, airflow.

Indirect Costs. All of the costs which cannot be directly attributed in their entirety to the individual unit of production or project.

Indirect Luminaire. A lighting fixture that distributes most of its light upward.

Indirect Overhead. A firm's general office overhead expenses that are not attributable to any specific job and that would continue even when there is no work on hand.

Indirect System. A solar heating or cooling system in which the solar heat is collected outside the building and transferred inside using ducts or piping and fans or pumps.

Indirect Waste Pipe. An indirect waste pipe is a pipe that does not connect directly with the drainage system but conveys liquid wastes by discharging into a plumbing fixture, interceptor, or receptacle which is directly connected to the drainage system.

Individual High Chairs. A welded wire bar support used under a support bar, to provide support for top bars in slabs, joists, or beams; also used to support upper mats of bars in slabs without support bars.

Individual Vent. A pipe which is installed to vent a fixture trap and which connects with the vent system above the fixture served or terminates in the open air.

Indoor Meter Center. A room or area inside a building or structure where the meters for measuring the usage of water, electricity, or gas are located.

Induced Draft. In a boiler, combustion air drawn through the burner or fuel bed by a power driven fan in the flue.

Induced Magnetism. Ability of a magnetic field to produce magnetism in a metal.

Inductance. A property of an electric circuit by which an electromotive force is induced in it by a variation of current either in the circuit itself or in a neighboring circuit.

Induction Baking. Using heat induced by electrostatic and electromagnetic means for baking of finishes.

Induction Brazing. A brazing process wherein coalescence is produced by the heat obtained from resistance of the work to the flow of induced electric current and by using a nonferrous filler metal having a melting point above 800° F. but below that of the base metals; the filler metal is distributed in the joint by capillary attraction.

Induction Motor. An a-c motor which operates on principle of rotating magnetic field; rotor has no electrical connection, but receives electrical energy by transformer action from field windings.

Induction Welding. A welding process wherein coalescence is produced by the heat obtained from resistance of the work to the flow of induced electric current with or without the application of pressure.

Inductive Reactance. Electromagnetic induction in a circuit creates a counter or reverse (counter) emf (voltage) as the original current changes; it opposes the flow of alternating current.

Industrial Compactor. A machine in an industrial setting that compacts or compresses materials.

Industrial Equipment. Mechanical or non-mechanical devices used in an industrial setting.

Industrial Fluorescent. A large fluorescent light fixture used in an industrial setting.

Industrial Hoist. A mechanical device used to lift loads in an industrial setting.

Industrial Hygienist. A professional qualified by education, training, and experience to anticipate, recognize, evaluate, and develop controls for occupational health hazards.

Industrial Wastes. Liquid wastes which result from the processes employed in industrial establishments and which are free of fecal matter.

Industrial Wood Floor. A heavy duty wood floor made of 2 inch thick decking or of wooden blocks laid on end; very resistant to heavy loads and traffic; also called a Factory Floor.

Inelastic Analysis. An analysis of deformations and internal forces based on equilibrium, nonlinear stress-strain relations for concrete and reinforcement, consideration of cracking and time dependent effects and compatibility of strains; the analysis shall represent to suitable approximation the three dimensional action of the shell together with its auxiliary members.

Inelastic Behavior. Behavior of an element beyond its elastic limit.

Inert. Chemically unreactive.

Inert Gas. 1. A gas that is not reactive, such as helium or neon; see Noble Gas. 2. In welding, a gas which does not normally combine chemically with the base metal or filler metal.

Inertia. A property of matter by which it continues in its present state of rest or uniform motion in a straight line, unless that state is changed by an external force.

Inertia Block. A concrete pad on which to mount mechanical equipment on vibration isolators, to prevent machinery vibration from being transmitted into the building structure.

Inert Pigment. A nonreactive pigment, filler.

Infiltration. The exchange between conditioned room air and outdoor air through cracks and openings in the building enclosure.

Inflammable. Capable of being easily ignited; flammable; the opposite is nonflammable.

Inflection Point. A point on a curve at which the curvature is zero; in the elastic curve it is the point of zero moment; usually, the curvature of the curve lying to one side of the inflection point is positive, while to the other side it is negative.

Information Technology. The use of machines to handle, process, and transmit information.

Infrangible. Not capable of being broken or separated into parts.

Infrared. Lying beyond the visible spectrum at its red end.

Infrared Lamp. Electrical device which emits infrared rays; invisible rays just beyond red in the visible spectrum.

Infra-Red Sensor. A mechanical device that detects motion using infra-red radiation.

Inglaze Decoration. A ceramic decoration applied on the surface of an unfired glaze and matured with the glaze.

Ingot. An oblong piece of cast metal.

Ingress. Entrance; the right to enter.

Inhibitive Pigment. One which retards the corrosion process.

Inhibitor. A substance which prevents chemical reaction such as corrosion or oxidation.

Initial Flow. See Flow of Mortar.

Initial Set. The start of the setting action in mortar, grout or concrete; see Set,4.

Initial Setting Time. The time required for a freshly mixed cement paste, mortar or concrete to achieve initial set.

Initial Stress. In prestressed concrete, the stresses occurring in the prestressed members before any losses occur.

Injection Molding. A method of manufacturing plastic objects by injecting thermoplastics into molds.

Injunction. A court order that requires a party to do or not to do a certain act.

Inlay. A decoration in which the design is set into the surface.

Inlet. An opening for intake.

Inlet Fitting. A fitting or fixture through which circulated or hydro jetted water enters the pool, spa or hot tub.

Inlet or Pool Return. Fitting through which filtered water flows into pool.

Innovation. The introduction of something new.

Inorganic. Not organic; usually of mineral origin.

Inorganic Coatings. Chemical colors obtained by combining two or more inorganic chemicals.

Insect Screen. A wood or metal frame with fine wire netting to keep out insects, yet permits ventilation.

Insert. A device built into the concrete formwork so it is cast integrally into the concrete, usually for attaching something later.

Inside Corner Trowel. A finishing device used by drywallers and plasterers to finish an inside corner with one pass of a trowel.

Inside Drain. A roof drain that is not at the perimeter of the roof, requiring piping inside the building to carry roof drainage to the exterior.

Insolation. The total amount of solar radiation — direct, diffuse and reflected — striking a surface exposed to the sky.

Insolvency. The financial situation in which liabilities exceed assets.

Inspection. Examination of completed work or work in process to determine its conformance to the contract requirements and the building code.

Inspection Hole. A temporary or permanent opening in work to facilitate inspection.

Inspection List. List of defective or incomplete work to be completed after substantial completion; also called Punch List.

Inspector. One who examines completed work or the work under progress, such as a building inspector.

Instructions to Bidders. Instructions contained in the bidding documents for preparing and submitting bids for a construction project.

Instrument. Used broadly to denote a device that has measuring, recording, indicating or controlling abilities.

Insulated. Covered and protected in an approved manner with suitable materials of the type and thickness adopted as standard requirements for the voltage and location of the particular material, conductor or part involved.

Insulated Block. Hollow masonry block filled with insulation.

Insulated Glass. A type of glass constructed in a manner to protect against sound, heat, heat loss, or moisture; double or triple glazed glass.

Insulated Roof. The roof of a building, constructed of materials that protect against heat, heat loss, or moisture.

Insulated Siding. Wall siding that has been manufactured with insulation installed under the finish.

Insulating Varnish. A varnish especially designed for the electrical insulation of wires, coils, and electrical appliances.

Insulating Wallboard. Foil Backed Gypsum products; see Foil Back.

Insulation. Any material used to reduce the effects of heat, cold or sound transmission and to reduce fire hazard; any material used in the prevention of the transfer of electricity, heat, cold, moisture and sound.

Insulation Board. 1. Insulation of various types in the form of a rigid board. 2. A low-density board made of wood, sugar cane, cornstalks, or similar material; it is dried and usually pressed to a thickness of 1/2 or 25/32 inch; also called Fiberboard.

Insulation, Calcium. See Calcium Insulation.

Insulation, Duct. See Duct Insulation.

Insulation, Electric. See Electric Insulation.

Insulation Equipment. Mechanical devices used to distribute insulated materials in walls, attics or crawlspaces.

Insulation Fastener. A device used to fasten different types of insulation to floors, walls, roofs, or ceiling surfaces.

Insulation Fasteners. Devices used to fasten different types of insulation to floors, walls, roofs, or ceiling surfaces.

Insulation, Gymnasium Underfloor. See Gymnasium Underfloor Insulation.

Insulation, Masonry. See Masonry Insulation.

Insulation Material. Material which is a poor conductor of heat; used to retard or slow down flow of heat through wall or partition.

Insulation, Pipe. See Pipe Insulation.

Insulation Removal. The process of ridding a building or structure of insulation.

Insulation, Rigid. See Rigid Insulation.

Insulation, Sill Sealer. See Sill Sealer Insulation.

Insulation, Sound Attenuating. See Sound Attenuating Insulation.

Insulation, Sprayed. See Sprayed Insulation.

Insulative Panel, Foil-Face. Flat sheet material made of insulative material that has aluminum foil on one face to improve thermal resistance characteristics of the assembly.

Insulative Panel. Flat sheet material made of insulative material to improve thermal resistance characteristics of assembly.

Insulative Wall Sheathing, Foil-Faced. Flat sheet material of foil-faced insulative board material secured to the exterior side of exterior wall studs, used to create a rigid surface on the building superstructure and serve as base to receive siding or veneer construction.

Insulator. A ceramic fitting which insulates electrical wires.

Insulator Spool. A device designed and used to support a conductor and electrically separate it from other conductors or objects.

Insurance. A contract whereby one party undertakes to indemnify or guarantee another against loss from specified perils.

Insurance Premium. The amount paid to an insurance company for protection against property or business losses and liabilities.

Insured. Covered by insurance.

Intaglio. A carving incised in hard material.

Intangible. Something that cannot be precisely measured or assessed; impalpable; not having physical substance.

Intangible Assets. Assets having no physical nature, but having value derived from the rights conferred upon their owner, such as goodwill.

Integral Cove Base. A coved base produced by running a flooring material up the wall.

Integrated System. A solar heating or cooling system in which the solar heat is absorbed in the walls or roof of a dwelling and flows to the rooms without the aid of complex piping or ducts.

Intense Color. A strong, vivid color.

Intensity. A subjective measure of the force of an earthquake at a particular place as determined by its effects on persons, structures and earth materials; intensity is a measure of effects as contrasted with magnitude which is a measure of energy.

Intensity Level (IL). A measure of the acoustic power passing through a unit area expressed on a decimal scale.

Intercepting Drain. A drain that intercepts and diverts ground water before it reaches the area to be protected; also called a Curtain Drain.

Interceptor. Any device installed in the drainage system to prevent the passage of grease or other solid materials such as sand.

Intercoat Contamination. In painting, the presence of foreign matter between successive coats.

Intercom. An Intercommunication System.

Intercommunication System. A local telephone system within a building or a group of rooms or offices.

Interest. A charge for borrowed money; to calculate the amount of interest. Interest = Principal x Rate x Time.

Interest Rate. The amount of interest, expressed as a percentage of the principal per period of time, for example, 8 percent per annum.

Interface. 1. A surface forming a common boundary between two regions. 2. A point where interaction occurs between two systems or processes. 3. An electrical component that connects a computer to another piece of equipment.

Interference Body Steel Bolt. High strength bolt with hardened steel ribs on bolts shank to facilitate driving of bolts in misaligned holes.

Intergrown Knot. A knot whose rings of annual growth are completely intergrown with those of the surrounding wood.

Interim Financing. A construction loan, to be paid off from the permanent loan upon completion of construction; also called Interim Loan.

Interim Loan. See Interim Financing.

Interior Decorator. One who selects, and sometimes purchases, furniture, floor coverings, and fabrics, and designs interior color schemes.

Interior Designer. One who designs interiors of buildings.

Interior Elevation. An architectural drawing showing the projection on a vertical plane of an interior surface of a building.

Interior Finish. Material used to cover the interior framed areas or materials for walls and ceilings.

Interior Painting. The process of painting the inside surfaces and trim of a building or structure.

Interior Plywood. A general term for plywood manufactured for indoor use or in construction subjected to only temporary moisture; the adhesive may be interior, intermediate or exterior.

Interior Stucco. Finish plaster for walls and ceilings, finishing, smooth, or textured, consisting of a mechanically blended compound of Keene's cement, lime (Type S), and inert fine aggregate; color pigment may be added to produce integrally colored interior stucco; also called Interior Plaster; see Stucco.

Interior Trim. General term for all the molding, casing, baseboard, and other trim items applied within the building by finish carpenters.

Interior Veneer. A veneer applied to surfaces other than weather-exposed surfaces.

Interior Wood Paneling. Wood panel members used to cover and finish interior walls; often prefinished.

Interior Wood Trim. Finish components of wood such as moldings applied around openings and intersections of surfaces at interior locations.

Interlocked. So arranged that a sequence of operation or steps are electrically or mechanically assured.

Interlocked-Grained Wood. Grain in which the fibers put on for several years may slope in a right-handed direction, and then for a number of years the slope reverses to a left-handed direction, and later changes back to a right-handed pitch, and so on; such wood is exceedingly difficult to split radially, though tangentially it may split fairly easily.

Interlocking. The tiering, binding or blending of various articles with one another.

Intermediate Coat. Middle coat; guide coat.

Intermittent. Intermittent duty is a requirement of service that demands operation for alternate intervals of (1) load and no load; or (2) load and rest; or (3) load, no load and rest.

Intermittent Cycle. Cycle which repeats itself at varying time intervals.

Intermittent Welding. Welding wherein the continuity is broken by recurring unwelded spaces.

Internal Control. The method and procedures adopted within a business to safeguard assets and control its operations.

Internal Friction. Resistance to sliding within a mass of wet or dry earth, sand, or gravel.

Internal Mix. A spray gun in which the fluid and air are combined before they leave the gun.

Internal Vibrator. Metal vibrating head immersed in fresh concrete during placement to consolidate fresh concrete; used primarily in cast in place construction.

International Building Code. Produced by International Code Council.

International Symbol of Accessibility. A symbol adopted by Rehabilitation International's 11th World Congress for the purpose of indicating that buildings and facilities are accessible to persons with disabilities.

Interpass Temperature. In a multiple-pass weld, the lowest temperature of the deposited weld metal before the next pass is started.

Interply. Between two layers of roofing.

Interpolate. Estimate the in between values from known ones in the same range.

Interrogatories. Written questions that a person in a lawsuit or litigation must answer in writing, under oath.

Interrupting Capacity. The maximum current at its rated voltage that a device will open without injury to any part other than replaceable fuses. The interrupting capacity shall be determined from the recognized standards for testing equipment.

Intersect. To met and cross at a point.

Intersection. A place where two or more things meet and cross at a point.

Interstate Commerce Commission. Government body which controls the design and construction of pressure containers.

Interstice. A space that is between things, like the grout space between ceramic tiles.

Intestate. A person not having made a will before death.

In-the-White. Natural or unpainted; the natural unfinished surface of the wood.

Intumesce. To foam, swell, froth, or bubble up as a result of heat, liquid, air or chemical action.

Intumescence. An enlarging, swelling or bubbling up, as under the action of heat.

Inventory. Includes all material, labor and overhead on jobs which are currently in progress, less any amounts which have been billed to customers.

Invert of a Pipe. The inside bottom elevation; the flow line.

Invert, Manhole. The lowest inside surface of a manhole; a channel in the manhole through which wastewater or stormwater flows.

Invert. The lowest portion of the inside of any horizontal pipe.

Inverted Arches. Arches that appear to have been built upside down; used to distribute the weight of the wall from pier to pier.

Inverted Roof. A membrane roof assembly in which the thermal insulation lies above the membrane.

Inverted Roof Membrane Assembly (IRMA). A roofing system where the membrane is applied to the substrate and the insulation boards are on top, weighted down by gravel and rock ballast.

Invisible Hinge. A hinge so constructed that no parts are exposed when the door is closed.

Invitation to Bid. A formal written invitation to submit a bid, usually placed in trade papers or newspapers, informing prospective bidders about a project; necessary for public work but not for private work.

Invited Bidders. A closed list of bidders decided by owner and architect as the only ones who may bid a job.

Invoice. A list of goods shipped or services rendered, with prices and charges.

Involuntary Lien. Any legal claim that is placed on a property without the permission of the owner, such as a mechanics lien.

Involute Curve. A curve traced by a thread unwound from another curve.

Iodine. A chemical compound containing iodine; potassium or sodium iodide, when used with a suitable oxidizing agent such as chlorine, will release iodine in pool water.

Iodine Number. A means of identifying and specifying qualities of oils, resins and waxes, based on fact that different qualities of these products will absorb different quantities of iodine.

Ion. Group of atoms or an atom electrically charged.

Ionic. The order of Greek architecture characterized by a column with scroll shapes on either side of the capital.

Ionization. Breaking up of molecules into two or more oppositely charged ions; an ion is one of electrified particles into which molecules are divided by the use of water and other solvents.

Ionization Detector. A mechanical device indicating the presence of certain gaseous compounds and amount of ions produced when subjected to ultraviolet light or hydrogen flame.

IPM. Integrated Pest Management.

IPS. Iron pipe size

IR Drop. Electrical term indicating the loss in a circuit expressed in amperes times resistance (1 x R) or voltage drop.

Irish Confetti. A rock or brick used as a missile.

IRMA. Inverted Roof Membrane Assembly.

Iron. A heavy malleable ductile magnetic silver-white metallic element that readily rusts in moist air and is the most used of metals.

Iron Blue. Blue pigment which depends on iron content to provide blue color.

Iron Driers. Driers with high tinting strength which limits use to colored finishes.

Iron Oxide. Ferrous oxide; corrosion of iron by oxidation of the surface; rust; iron oxide is available in three forms; red, brown and yellow; it is known under a variety of names, such as Red Oxide, Jeweler's Rouge, Venetian Red, Ferric Oxide, Indian Red, Red Ochre, Mineral Rouge, Spanish Oxide, and Turkey Red.

Iron Phosphate Coating. Conversion coating; chemical deposit.

Iron-Phosphate Treated. Cleaning and treatment of steel deck prior to forming.

Ironstone Ware. Stone china, white granite ware, historic terms for a durable English earthenware.

Irregular Stone. Stone cut to or quarried in various shapes and sizes.

Irrigation, Lawn. To supply water to grassy areas by artificial means.

Isinglass. Mica.

Isobar. Lines on a map connecting points of equal atmospheric pressure at a particular time.

Isochronous. Occurring at the same time.

Isocyanate Resins. Urethane resins.

Isogram. A line on a map or chart connecting points of equal value, as of temperature, elevation, or precipitation.

Isohyetal Line. A line on a map or chart connecting points of equal rainfall in a specified period.

Isolated. Not readily accessible to persons unless special means for access are used.

Isolation. An acoustical term used to describe sound privacy by reducing direct sound paths.

Isolation Interface. The boundary between the upper portion of a structure, which is isolated, and the lower portion of the structure, which moves rigidly with the ground.

Isolation Joint. A separation between adjoining parts of a concrete structure, usually a vertical plane, at a designed location such as to interfere least with performance of the structure, yet such as to allow relative movement and avoid formation of cracks elsewhere in the concrete and through which all or part of the bonded reinforcement is interrupted.

Isolation System. The collection of structural elements which includes all individual isolator units, all structural elements which transfer force between elements of the isolation system, and all connections to other structural elements; the isolation system also includes the wind restraint system if such a system is used to meet the design requirements.

Isolator Unit. A horizontally flexible and vertically stiff structural element of the isolation system which permits large lateral deformations under design seismic load; an isolator unit may be used either as part of or in addition to the weight-supporting system of the building.

Isolator, Vibration. See Vibration Isolator.

Isometric Drawing. The representation of an object in isometric projection; an isometric drawing looks like a perspective but all lines parallel to the three major axes are measurable.

Isometric Projection. Axonometric projection in which all three faces are equally inclined to the drawing surface.

Isosceles Triangle. A triangle having two equal sides.

Isoseismals. Map contours drawn to define limits of estimated intensity of shaking for a given earthquake.

Isothermal. Changes of volume or pressure under conditions of constant temperature.

Isothermal Expansion and Contraction. Action which takes place without a temperature change.

Isotropic. Exhibiting the same properties in all directions.

Ivory. A hard creamy white substance that makes up the tusk of an elephant, walrus, hippopotamus, or narwhal.

Ivory Black. A high grade bone black pigment; formerly made by charring or burning ivory.

Ivory Paper. See Cream Paper.

Ivory Tower. 1. A state of seclusion or separation from the practicalities and realities of life. 2. Academia.

J

Jack. A mechanical device for lifting heavy objects.

Jack Arch. 1. One having horizontal or nearly horizontal upper and lower surfaces; also called a Flat Arch or Straight Arch. 2. Any arch that is roughly built.

Jacket, Aluminum. See Aluminum Jacket.

Jackhammer. A pneumatic concrete breaker.

Jacking Equipment. In prestressed concrete, the device used to stress the tendons.

Jacking Force. The temporary force exerted by a device that introduces tension into prestressing tendons in prestressed concrete.

Jacking Pipe. Forcing pipe through the ground in a tunnel created by the pipe itself; the pipe is generally jacked horizontally in short lengths.

Jacking Stress. In prestress concrete, the maximum stress occurring in a tendon during stressing.

Jack-of-all-Trades. A person who can do many different kinds of work.

Jack Plane. A medium sized plane used in the first stages of planing.

Jack Rafter. 1. A short rafter that joins a hip rafter to the top of the wall plate. 2. A short rafter that joins a valley rafter to a ridge board.

Jacquard. Mechanism for a Wilton carpet loom which produces the desired color design using a chain of perforated cardboard cards punched according to the design elements, which when brought into position activates this mechanism by causing it to select the desired color of yarn to form the design on the pile surface; the unselected colors are woven dormant through the body of the fabric.

Jagged Edges. Irregularities left on the edges of ceramic tile due to the use mainly of hand cutting tools.

Jalousie. A blind or shutter made of a row of angled slats to keep out rain and control the influx of light.

Jalousie Window. A window with stationary or adjustable blinds angled to permit air and provide shade, while preventing rain from entering.

Jamb. The vertical side of a door or window frame.

Jamb Anchor. Steel anchor used to fasten a steel door frame to the wall or partition construction.

Jamb Block. A concrete masonry unit with a preformed slot to receive a window or door frame.

Jamb, Door. See Door Jamb.

Japan Color. Colored paste made by grinding high-quality colors in hard drying varnish.

Japan Drier. Varnish gum with a large proportion of metallic salts, such as lead, cobalt, or manganese, added to hasten drying; used in paints, varnishes, and enamels.

Japanese Lacquer. Varnish made from sap of a tree which grows in Japan; becomes very hard and black as it dries.

Jaspe. Carpet surface characterized by irregular stripes produced by varying textures or shades of the same color.

J-Box. An electrical junction box.

Jib. The arm of a crane.

Jib Crane. A crane which has a projecting arm off its derrick boom.

Jig. A tool or template used to maintain mechanically the correct positional relationship between a piece of work and the tool or between parts of work during assembly.

Jiggering. Forming ceramic ware from a plastic body by differential rotation of a profile tool and mold, the mold having the contour of one surface of the ware and the profile tool that of the other surface.

Jigsaw. 1A machine saw with a narrow reciprocating blade used for cutting curved and intricate patterns. 2. A scroll saw.

Jitterbug. A grate tamper for pushing coarse aggregate slightly below the surface of a slab to facilitate finishing.

J-Mold. A metal molding strip with a J-like section, used to edge plaster or gypsum wallboard.

Job. A piece of work.

Job Built Form. A temporary structure or mold constructed on a jobsite, for the support of concrete while it is setting and gaining sufficient strength to be self-supporting.

Job Lot. A miscellaneous collection of goods for sale as a lot.

Job Requirements. A list of specific, necessary, and essential tasks to bring to completion a building or structure; also called General Requirements.

Job Scheduling. An itemization in chronological order, often in chart form, of project tasks in order to start and complete a building or structure.

Jobsite. The place where construction takes place.

Job Trailer. A towed vehicle placed on a jobsite, acting as an office space.

Joinery. The art or trade of joinery; woodwork; finish carpentry.

Joining. The juncture of two separate plaster applications usually within a single surface plane; also called a Jointing.

Joint. 1. The location where two or more members are to be joined. 2. The point of connection between structural members. 3. The seam produced by the placement of two pieces of gypsum board together but not necessarily in the same plane. 4. The junction of two pieces, as of wood or veneer. 5. The location at which two pieces of pipe or a pipe and a fitting are connected together.

Joint, Butt. See Butt Joint

Joint Cement. See Joint Compound.

Joint Check. A check made payable to more than one payee.

Joint Clip. 1. A metal fastener used vertically, sharp edge down, over the edges of two pieces of wood, and then hammered down into them. 2. In plywood sheathing, the clip fastens two abutting pieces of plywood.

Joint Compound. A cementitious material used in covering joints, corners and fasteners in the finishing of gypsum board to produce a smooth monolithic surface; also called Spackle or Joint Cement.

Joint, Contraction. See Contraction Joint.

Joint, Control. See Control Joint.

Joint Control. An independent escrow used to safeguard and disburse construction funds.

Joint Efficiency. The strength of a wood joint expressed as a percentage of the strength of clear straight-grained material.

Jointer. 1. A metal hand tool used to cut a joint partly through fresh concrete. 2. In masonry, a tool used by bricklayers to form the various types of mortar joints between the courses of masonry, as the V, the concave, and weather joints. 3. An electrically powered machine for making joints.

Joint, Expansion. See Expansion Joint.

Joint Factor. Joint efficiency.

Joint Filler. Compressible material used to fill a joint to prevent the infiltration of debris and to provide support for sealants.

Jointing. 1. Smoothing and straightening the edge of a wood board. 2. Jointing is done automatically by a jointer. 2. Grinding or filing the teeth or knives of power tools to the correct height; circular saws are jointed so that there are no high or low teeth; knives of planers and jointers are jointed so that each knife makes the same depth of cut as all others. 3. The act of facing the mortar joints.

Joint Photographing. In gypsumboard, the shadowing of the finished joint areas through the surface decoration.

Joint Reinforcement. Steel wire, bar, or prefabricated reinforcement which is placed in mortar bed joints.

Joint Reinforcing Mesh. A woven fiber screen-like material used in lieu of paper joint tape.

Joint Reinforcing Metal. Strips of expanded metal, woven or welded wire mesh used to reinforce corners and other areas of plaster and lath.

Joint Reinforcing Tape. A type of paper, metal, fabric, glass mesh, or other material, commonly used with a cementitious compound, to reinforce the joints between adjacent gypsum boards.

Joint Reinforcing. Joint Reinforcement.

Joint Sealant. Compressible material used to exclude water and solid foreign materials from openings; a substance that prevents water and moisture from entering joints; also called Joint Sealer.

Joint Sealer. Joint Sealant.

Joint Tape System. The compound and tape system used to conceal and finish joints in gypsum board.

Joint Tape. Paper tape that is applied with gypsum-based joint compound to treat the joints in gypsum wallboard.

Joint Tenants. Two or more persons owning property together with the right of survivorship.

Joint Venture. A temporary partnership composed of individuals, partnerships, or corporations organized to accomplish a specific job or course of work.

Joist. A, usually horizontal, structural member used as a floor, ceiling or roof framing member.

Joist Anchor. A metal rod incorporated into a masonry wall that extends out to be fastened to a joist or rafter.

Joist Bridging. 1. The bracing of joists by the fixing of lateral members between them. 2. Pieces fitted in pairs from the bottom of one floor joist to the top of adjacent joists, and crossed to distribute the floor load. 3. Wood pieces of width equal to the joists and fitted neatly between them. 3. Diagonal or longitudinal members used to keep horizontal members properly spaced, in lateral position, vertically plumb, and to distribute load.

Joist Girder. A light steel truss used to support open-web steel joists.

Joist Hanger. A metal stirrup that supports the ends of joists so that they are flush with the girder.

Joist, Metal. See Metal Joist

Joist Painting. The act or process of applying paint to seal, protect, or add color to a joist, usually refers to metal joists.

Joist, Sister. See Sister Joist.

Joist, Wood. See Wood Joist.

Joule. In the international system of units, the amount of energy needed to raise the temperature of 1 kilogram of water 1 degree Celsius.

Joule-Thomson Effect. The change in the temperature of a gas on its expansion through a porous plug from a higher pressure to a lower pressure.

Journal. 1. A record of current transactions. 2. A log. 3. The part of a rotating shaft, axle, roll, or spindle that turns in a bearing.

Journal, Crankshaft. Part of shaft which contacts the bearing on the large end of the piston rod.

Journeyman. 1. A fully trained and qualified person in a craft or trade such as a carpenter, plumber, or electrician. 2. A plasterer or lather who through training and experience has become thoroughly skilled in his trade; distinguished from an apprentice or a laborer. 3. A painter who has had at least three year's experience and schooling as an apprentice.

Judgment. A final decision of a court.

Jumbo Brick. A brick that is larger than standard size and measures 8"x4"x4".

Jumpover. In piping, a double offset used to return the pipe to its original line; also called a Return Off-Set.

Junction Box. A metal or hard plastic electrical rough-in box, usually square or octagonal, housing only wire or cable connections.

Jute. Strong, durable yarn spun from fibers of the jute plant, native to India and Far East, used in the backings of many carpets.

Jute Padding. A padding made of a durable yarn that comes from plant fiber used as an underlayment for carpet.

Juxtaposition. The placing of things side by side.

Juxtaposition of Colors. Placing colors side-by-side, or close together; complementary colors such as blue and orange in juxtaposition accentuate each other.

K

Kalotermes. See Termite.

Kalsomine. See Calcimine.

Kaolin. 1. A refractory clay consisting essentially of minerals of the kaolin group and which fires to a white or nearly white color; also called China Clay. 2. Inert pigment which tends to impart easy brushing properties to paint products in which it is used.

Kata Thermometer. Large-bulb alcohol thermometer used to measure air speed or atmospheric conditions by means of cooling effect.

Kauri Gum. A fossil copal found in New Zealand.

Kauri Reduction. Test for solvent power of petroleum solvents.

K Bracing. That form of bracing where a pair of braces located on one side of a column terminates at a single point within the clear column height.

Keene's Cement. A cement composed of finely ground, anhydrous, calcined gypsum, the set of which is accelerated by the addition of other materials, used in areas subjected to moisture; a hard, strong finishing plaster that is made from gypsum and maintains a high polish; because of its density, it excels for use in bathrooms and kitchens and is also widely used for the finish coat in auditoriums, public buildings, and other places where walls may be subjected to unusually hard wear or abuse; also called Anhydrous Calcined Gypsum.

Keeper. See Strike, 2.

Kelly Ball Test. A method by which the workability of fresh concrete can be determined.

Kelvin Scale. A temperature scale on which the unit of measurement equals the Celsius degree and according to which absolute zero is 0°, the equivalent of minus 273.16° C; water freezes at 273.16° K. and boils at 373.16° K.

Kennel. 1. A shelter for a dog. 2. A breeding or boarding establishment for dogs.

Kerf. 1. A cut made with a saw, or with a cutter, part way through a material, generally to facilitate breaking to a desired shape. 2. The space from which metal has been removed by a cutting process.

Kerfing. Longitudinal saw cuts or grooves of varying depths on the unexposed faces of millwork members to relieve stress, prevent warping, or to facilitate bending.

Kerosene. A distillate obtained in petroleum refining which evaporates slowly.

Ketones. Organic solvents containing CO grouping; commonly used ketones are acetone (dimethyl ketone); MEK (methyl ethyl ketone); and MIBK (methyl isobutyl ketone).

Kettle. Portable roofer's equipment for heating bitumen to a suitable temperature for use.

Kettle Bodied Oil. Oil which has been held at an elevated temperature until the oil has thickened.

Kettle, Steam. See Steam Kettle.

Key. 1. A slot formed into a concrete surface for the purpose of interlocking with a subsequent pour of concrete; a slot at the edge of a precast member into which grout will be poured to lock it to an adjacent member. 2. A small piece of wood inserted in one or both parts of a joint to align it and hold it firmly together. 3. The center stone or brick of an arch. 4. The grip or mechanical bond of one coat of plaster to another coat or to a plaster base. It may be accomplished physically by the penetration of wet mortar or crystals into paper fibers, perforations, scoring irregularities, or by the embedment of the lath; see Mechanical Bond.

Key Changes. The different combinations of bittings or tumbler arrangements in a series of locks.

Keyed Joint. 1. A joint in which one structural member is keyed or notched into an adjoining member as in timber construction. 2. In masonry construction, a finished joint of mortar which has been tooled concave.

Keyhole Saw. A narrow pointed fine tooth saw for cutting small or curved holes.

Key Lock. A lock which is activated by a key.

Keystone. The wedge shaped central stone at the top of an arch.

Key Switch. An electric on-off switch operated by a removable key.

Kick Hole. A hole in the roofing membrane at the base of a parapet wall, at the cant, usually caused by being stepped on or kicked.

Kick Lift. A jacking wedge used to elevate or shift the gypsum board into proper nailing position on the wall during the application procedure.

Kick Plate. A metal plate or strip that runs along the bottom edge of a door to protect against the marring of the finished surface.

Kiln. A furnace, oven, or heated enclosure for processing a substance by burning, firing, or drying.

Kiln Dried Lumber. Lumber that has been kiln-dried, often to a moisture content of 6 to 12%; common varieties of softwood lumber, such as framing lumber, are dried to a somewhat higher moisture content.

Kiln Dried. Lumber that has been heated in a kiln to dry and control the amounts of moisture.

Kiln Drying. Drying of wood, paint, varnish or lacquer in room or compartment with heat and humidity regulated.

Kiln Run. Brick or structural clay tile from one kiln which have not been sorted or graded for size, burning, or color variation.

Kilogram. A metric unit of mass and weight equal to 1000 grams.

Kilometer. A metric unit of distance equal to 1000 meters.

Kilowatt. An electrical unit of 1000 watts.

Kilowatt-Hour (KWH). A measure of electrical energy equivalent to a power consumption of 1000 watts for one hour.

Kinematic Viscosity. A method of measuring viscosity using the stoke as the basic measurement unit.

Kinetic. Relating to motion.

Kinetic Energy. Energy associated with motion.

Kinetic Friction. The friction between two surfaces that are moving tangentially in contact with each other; kinetic friction is always slightly less than static friction.

King. See King Closer.

King Closer. In masonry, a closer used to fill an opening in a course larger than a half brick; about three- fourths the size of a regular sized brick; also called a king.

King Post. In a roof system, the member placed vertically between the center of the horizontal tie beam at the lower end of the rafters and the ridge, or apex of the inclined rafters.

King Post Truss. A standard peaked roof truss configuration.

King Stud. The last full length wall stud adjacent to a trimmer stud at a door or window opening.

King Valve. Liquid receiver service valve.

Kip. A unit of weight or force equal to 1,000 pounds.

Kirk. Church.

Kitchen. A room for preparation of food.

Kitchen Cabinet. A case, box, or piece of furniture with sets of drawers or shelves, with doors, primarily used for storage, mounted on walls or floors in a kitchen area.

Kitchen Compactor. A machine in a kitchen that compresses or compacts materials by using hydraulic weight, force, or vibration.

Kitchen Faucet. A fixture found in the kitchen that is used for drawing potable water.

Kitchen Fryer. An appliance used for frying foods.

Kitchen Range. A stove with an oven and a flat top surface with burners powered by gas or electric current.

Kitchen Sink. A stationary basin that is found in the kitchen and connected to a drain.

Knee Brace. A corner brace, fastened at an angle from wall stud to rafter, stiffening a wood or steel frame to prevent angular movement.

Knee Kicker. A short tool with gripping teeth at one end and a padded cushion at the other, used in making small stretches during carpet laying.

Kneepads. Protective foam rubber pads for workers to wear strapped on their knees when working on floors.

Knee Wall. A short wall under the slope of a roof.

Knife. 1. A cutting implement consisting of a blade fastened to a handle. 2. A sharp cutting blade or tool in a machine.

Knife, Board. See Board Knife.

Knitted. A type of nonwoven carpet construction in which the backing yarn, stitching yarn and pile yarn are looped together with three sets of knitting needles, producing an uncut loop pile.

Knitted Carpet. Carpet made on a knitting machine by looping together backing, stitching and pile yarns with three sets of needles, as in hand knitting.

Knockout. A cylindrical removable section of a junction box or outlet box to accommodate installation of a conduit connector

Knot. That portion of a tree branch or limb that has been surrounded by subsequent growth of the stem; the shape of the knot as it appears on a cut surface depends on the angle of the cut relative to the long axis of the knot.

Knotty Pine. Pine wood whose knots are exposed, often used for cabinets and interior paneling.

Knotty Score. Of gypsum board, a ragged cut edge.

Knuckle. The enlarged part of a hinge into which the pin is inserted.

Knurled. Having a surface covered with small knobs or beads, as a nail which may have such a surface for greater holding power.

Kraftcord. Tightly twisted yarn made from wood pulp fiber, used as an alternate for cotton or jute in carpet backing.

Kraft Paper. A sturdy brown wrapping paper.

Krebs Unit (KU). Arbitrary units of viscosity.

K Series. A standard Steel Joist Institute designation for longspan steel joists.

KSI. Kips per square inch.

KU. Krebs Unit.

KWH. Kilowatt Hour.

Kyanite. The most abundant of the mineral polymorphs that include andalusite and sillimanite; $A_{12}O_3SiO_2$; used as a source of mullite in ceramics.

L

Labor and Material Bond. A bond, secured by the general contractor, which guarantees that the costs for labor and materials for the project will be paid.

Laboratory. A room or building equipped for physical or chemical testing, experimentation, or analysis.

Laboratory Cabinet. A case, box, or piece of furniture with sets of drawers or shelves, with doors, primarily used for storage, used in a place or area for scientific studies or commercial and institutional laboratories and testing facilities.

Laboratory Counter. A level surface in a laboratory where equipment is placed and kept and where work may be performed.

Laboratory Equipment. Devices and tools used in laboratory work.

Laboratory Hot Plate. A heating device in a laboratory.

Laboratory Incinerator. A device in a laboratory in which waste materials are burned.

Laboratory Incubator. An apparatus used in a laboratory for the maintenance of controlled conditions.

Laboratory Table. A surface on which laboratory materials and devices are kept and used.

Labor Union. A trade union.

Lac. A natural resin secreted by certain insects which live on the sap of trees in India and other Oriental countries; marketed in various forms, such as seed lac, button lac, and shellac.

Laches. Undue delay in asserting one's legal rights, resulting in loss of the rights.

Lacquer. A sometimes colored liquid made of shellac dissolved in alcohol, or of synthetic substances, that dries to form a hard protective coating.

Lacunar. A ceiling made up of vaults or recessed panels.

Lacunaria. A recessed panel in a ceiling.

Ladder. A frame consisting of two parallel side pieces connected by rungs at suitable distances to form steps on which persons may climb up or down.

Ladder Bar. A prefabricated reinforcement designed for embedment in the horizontal mortar joints of masonry; parallel deformed side rods connected in a single plane, by cross wires, thus forming a ladder-like design.

Ladder Cage. A cage-like structure surrounding a wall-mounted ladder for safety.

Ladder, Pool. See Pool Ladder.

Ladder Rung. A horizontal round member used as a step on a ladder.

Ladder Step. A horizontal flat member used as a step on a ladder.

Lag. Delay in response.

Lag Rod. A large diameter rod with a square or hexagonal head.

Lag Screw. A large diameter wood screw with a square or hexagonal head.

Laid Off. See Double-Up.

Laid-On. See Double-Up.

Laitance. A layer of weak and nondurable material containing cement and fines from aggregates, brought by bleeding water to the top of overwet concrete, the amount of which is generally increased by overworking or overmanipulating concrete at the surface by improper finishing or by job traffic; effluorescence.

Lake Asphalt. See Natural Asphalt.

Lake Pigment. Pigment made by putting an organic dye on a base of fine particles of inert or translucent pigment.

Lake Sand. Sand consisting predominantly of fine, rounded particles.

Lally Column. Tradename; a hollow steel column. sometimes filled with concrete.

Lamella. A unit of a surface network of closely spaced uniform ribs or beams, usually arranged in two or three intersecting diagonal lines; curved vaults and domes have been built of wood, steel, and concrete lamellas.

Lamina. The layers of material in a laminate.

Laminar Scale. Rust formation in heavy layers.

Laminate. 1. To form a product by bonding together two or more layers of materials. 2. The product so formed, such as a plastic laminate.

Laminated Glass. A glazing material consisting of outer layers of glass laminated to, and encasing, an inner layer of transparent plastic; used for automobile windshields and bulletproof glass.

Laminated Plastic. See Plastic Laminate.

Laminated Rubber. Several layers of rubber bonded together with adhesive under pressure.

Laminated Timber. An assembly made by bonding layers of veneer or lumber with an adhesive so that the grain of all laminations is essentially parallel.

Laminated Wallboard. Two or more layers of gypsum board held together with an adhesive.

Laminated Wood. A product made by bonding layers of veneer or lumber with an adhesive so that the grain of all layers is generally parallel.

Laminate, Paper-Base. A multilayer panel made by compressing sheets of resin-impregnated paper together into a solid mass.

Laminating Compound. A cementitious material, usually regular joint compound, used to adhere two or more layers of gypsumboard together.

Lamination. 1. The uniting of layers of material using adhesive. 2. The application of two or more layers of gypsum board.

Lampback. Pigment made by burning coal tar distillates without sufficient air; not quite true black.

Lampholder. A device to support an electric lamp mechanically and connect it electrically to the circuit conductors.

Lamp, Incandescent. See Incandescent Lamp

Lamp, Steri. See Steri Lamp.

Lanai. Porch, veranda, or covered patio.

Land. 1. The surface of the earth and all its natural resources. 2. A portion of the earth's solid surface defined by boundaries or ownership; privately or publicly owned. 3. See Root Face.

Land Contract. A contract for sale of land where title does not pass to the purchaser until all, or a certain number, of the payments have been made; also called Contract of Sale.

Landing. A platform between flights of stairs or at the termination of a flight of stairs.

Landing Terrazzo. A staircase landing with terrazzo tile mounted on its surface.

Landlord. The owner of property that is rented to a tenant.

Land Plaster. Coarsely ground natural gypsum used agriculturally as a soil conditioner.

Land Residual. A real estate appraisal technique where a reasonable return on the improvements is first deducted from the income, the balance being attributable to the land.

Landscape. 1. Natural scenery. 2. To improve a site by modification of the terrain, the planting of trees, shrubs, and ground cover, and the addition of hardscape.

Landscape Architect. One whose profession is designing the arrangement of land for human use involving vehicular and pedestrian ways and the planting of groundcover, plants, and trees.

Landscape Timber. Large, treated lengths of lumber used to decorate and act as soil erosion barriers and retaining wall members.

Land Surveyor. A person who surveys land.

Land Title. Porous clay pipe with open butt joints.

Lane Joint Cracks. Longitudinal separations along the seam between two paving lanes caused by a weak seam between adjoining spreads in the courses of the pavement.

Langly. A measure of solar radiation, equal to 1 calorie per square centimeter.

Lantern. A raised structure on a roof, glazed to admit light.

Lap. 1. The overlap of two roofing plies. 2. The length by which one bar or sheet of fabric reinforcement overlaps another. 3. The length of the overlap of two reinforcing bars; also called Lap Splice. 4. The amount of extension of one brick or any other masonry unit over another.

Lap joint. A joint made by placing one member partly over another and bonding the overlapped portions.

Lapped Joint. In wallpapering, a joint made by trimming one selvedge and overlapping the other.

Lapping. Smoothing a metal surface to high degree of refinement or accuracy using a fine abrasive.

Lap Siding. See Bevel Siding.

Lap Splice. See Lap, 3.

Large Calorie. See Calorie, 2.

Laser. 1. A device that generates an intense beam of coherent monochromatic radiation in the infrared, visible, or ultraviolet region of the electromagnetic spectrum, by stimulated emission photons from an excited source; used in communications, industry, and medicine; used in construction to establish accurate sightings, and plumb and level alignments, vertical and horizontal. 2. Acronym for Light Amplification by Stimulated Emission of Radiation.

Latch. 1.A bar with a catch and lever used as a fastening for a gate. 2. A spring-lock requiring a key to pass from the outside.

Latch Set. A fastening assembly for a door or window, operable from both sides, and lockable with a key.

Latent Ambiguity. A term of a contract that appears on its face to be unambiguous, but that is made ambiguous by external circumstances.

Latent Defect. A construction defect that is not perceptible by ordinary observation.

Latent Heat. Heat that changes the state of material from a solid to a liquid or a liquid to a gas.

Latent Heat Gain. The addition of heat to an enclosure by an increase in moisture content.

Latent Heat of Condensation. Amount of heat lost by a pound of a substance to change its state from a gas to a liquid.

Latent Heat of Vaporization. Amount of heat required per pound of a substance to change its state from a liquid to a gas.

Lateral. 1. Relating to the side. 2. At right angles to the long direction of the member; crosswise; transverse.

Lateral Brace. A temporary or permanent structural brace to resist lateral movement of a truss or beam.

Lateral Branch. See Horizontal Branch.

Lateral Force. A force acting generally in a horizontal direction, such as wind, earthquake, or soil pressure against a foundation wall.

Lateral Force Coefficients. Factors applied to the weight of a structure or its parts to determine lateral force for a seismic design.

Lateral-Force-Resisting System. The portion of the structure composed of members designed to resist forces related to earthquake, wind, or other lateral effects.

Lateral Load. 1. The horizontal component of the load produced by an arch, dome, vault, or rigid frame. 2. A horizontal load applied to a structure or member such as wind or earthquake.

Lateral Pressure. Horizontal pressure such as the force of soil against the side of a high foundation wall.

Lateral Slip Fault. A fault whose relative displacement is purely horizontal.

Lateral Support. A force or structural member that prevents a structure or earthen mass from moving in a lateral or horizontal direction.

Lateral Thrust. 1. The pressure that any load or force exerts sideways or through the ends of the members. 2. The horizontal component of the force or thrust produced by an arch, dome, vault, or rigid frame.

Latewood. The portion of the annual growth ring that is formed after the earlywood formation has ceased; it is usually denser and stronger mechanically than earlywood; also called Summerwood.

Latex. A water based emulsion of a synthetic rubber or plastic obtained by polymerization and used commonly in coatings and adhesives.

Latex Paint. A water-based paint that can be thinned and washed from applicators with water.

Latex-Portland Cement Grout. A portland cement grout with a special latex additive which results in a less rigid, less permeable grout than regular portland cement grout.

Latex-Portland Cement Mortar. A mixture of portland cement, sand, and special latex additives which is used for bonding tile to back-up material.

Lath. A wood strip, metal mesh, or gypsum board which acts as a backing and/or reinforcing agent for the plaster scratch coat or initial mortar coat; a plaster base.

Lath and Plaster Membrane. 1. A thin slab of lath and plaster including any integral supporting and stiffening members. 2. Lath and plaster as a unit of structure.

Lathe, Shop. See Shop Lathe.

Lath, Expanded Metal. See Expanded Metal Lath.

Lathe Tool. A blacksmith's tool used to handle hot metals.

Lath, Stucco. See Stucco Lath.

Lattice. A framework of crossed wood or metal strips.

Lattice Molding. Flat strip molding used in the construction of lattices.

Lattice Truss. A steel truss where the top and bottom chords are connected together by a steel lattice.

Lauan. A variety of Philippine mahogany, of moderate strength and durability, with wood of light yellow to reddish brown or brown.

Lauan Door. A door made of Philippine mahogany veneer.

Lauan Veneer. Very thin sheets of Philippine Mahogany, which may be combined with glue to create plywood or glued onto another surface to create a finished surface.

Laundry. A room for washing and drying of clothes and linens.

Laundry Chute. A chute that leads from an upper floor to a lower floor by which soiled laundry is conveyed.

Laundry Equipment. Appliances and materials used for cleaning clothes.

Laundry Tray. A tub with hot and cold water supply and a drain, used in a laundry room.

Lav. Lavatory.

Lavatory. 1. A basin with drainage and running water primarily used for washing the face and hands. 2. A room with a toilet and wash basin.

Lavatory Carrier. A horizontal structure attached to a bathroom wall to support and mount a lavatory.

Lavatory Faucet. A water dispensing device in a bathroom sink or lavatory.

Law. A rule of conduct enforced by courts.

Lawn. An area of ground covered with mown grass.

Lawn Irrigation. The supply of water to grassy areas by artificial means.

Lawsuit. A proceeding in which the jurisdiction of a court is invoked to resolve to a dispute between two or more parties.

Lawyer. A person who is educated in and authorized to practice law.

Layer. A stratum of weld metal consisting of one or more weld beads.

Laying Overhand. Laying brick on the farther face of a wall from a scaffold.

Lay Out. The measuring and setting out of work according to the design drawings.

Layout. A full-sized drawing showing arrangement and structural features.

Layout Stick. A long strip of wood marked at the appropriate joint intervals for the tile to be used; used to check the length, width, or height of the tilework; also called Idiot Stick.

Lazy Susan. 1. A revolving circular arrangement of shelves. 2. A circular revolving cabinet shelf used in corner kitchen cabinet unit.

Lb. Symbol for pound or pounds.

L Cut. A piece of tile cut or shaped like the letter L.

Leaching Field. A land area containing a series of subterranean perforated pipes that allow septic tank effluent to percolate into the soil; also called Disposal Field.

Leaching Pit. An excavated hole in the ground that can hold solids but allows liquids to pass through and leach into the ground.

Lead. 1. A heavy soft malleable ductile plastic but inelastic bluish white metallic element found mostly in combination, used in pipes, cable sheaths, batteries, solder, roofing, flashing, and shields against radioactivity. 2. A section of a wall built at the corners or other places wherever needed to act as a guide for the balance of the wall; the line of masonry work that is to be done is strung between the two leads which act as a guide.

Lead Carbonate, Basic. See Basic Lead Carbonate.

Lead Drier. Almost water-white drier which works on body of paint film; various combinations of lead, cobalt, and other driers are used in formulating many modern finishes.

Leaded Glass. A stained glass window, with the pieces of glass, often of irregular size and shape, set in a lead framework.

Leaded Zinc Oxide. White pigment made by combining lead sulphate and zinc oxide.

Leaded Zinc. Basic lead sulphate united with zinc oxide.

Leader. See Downspout.

Lead glass. A glass of high refractive index containing lead oxide.

Leadless Glaze. A ceramic coating matured to a glassy state on a formed article, or the material or the mixture from which the coating is made, to which no lead has been deliberately added; does not imply that the glaze is nontoxic or that it contains no lead; because of plant practices and conditions, a small percentage of lead, 0.1 to 0.2% (by dry weight), expressed as lead monoxide, may be present.

Lead Lined Wallboard. Gypsum wallboard with a ply of lead sheeting to be used in providing radiation protection in x-ray facilities.

Lead Oxide. Compound in several combinations of lead and oxygen, e.g., litharge and red lead.

Lead Sulphate, Basic. See Basic Lead Sulphate.

Lead Wool. A mass of lead shavings used primarily for packing or caulking.

Leaf. 1. The hinged or sliding part of a shutter, door, or gate. 2. One of a pair of doors.

Leafing. The overlapping arrangement of aluminum or gold bronze powders in a paint, similar to that of fallen leaves; good leafing is important in producing a metallic appearance and is caused by using treated or coated pigments along with suitable bronzing liquids.

Leak Detector. 1. Device or instrument such as halide torch, an electronic sniffer, or soap solution used to detect leaks. 2. Device used to detect and locate refrigerant leaks.

Leakage. 1. Entrance or escapage through a crack or hole, usually by a fault or mistake. 2. The loss of electricity through faulty insulation.

Lean Concrete. A concrete mix that is low in water and cement content, usually used as a stable base or fill.

Leaning Edge. In gypsumboard, a factory edge formed out of square with the surface.

Lean Mixture. Any plaster mortar containing a relatively high ratio of aggregate to cementitious material; a mortar mix that is too lean and has poor working qualities is said to be harsh.

Lean Mortar. Mortar containing a low percentage of cementitious components, not workable; see Harsh Mortar.

Lean-To. A shed on the side of a building with a sloping roof in a single plane.

Lease. An agreement whereby a tenant (lessee) obtains use of equipment, facilities, or real property for a specified period of time from the owner (lessor) usually by payment of a specified rent and under specified conditions.

Leaseback. A financial arrangement in which a tenant agrees to lease real property for a period during which the lease consideration will cover all mortgage payments and the lessor will own the property free and clear at the end of the lease term.

Lectern. A desk-like stand for holding a book or papers for a lecturer.

LED. Light-Emitting Diode.

Ledger. A wood or metal member bolted to a masonry wall to carry floor joists, ceiling joists, or rafters.

Ledger Bolt. A bolt set into a masonry wall to secure a wood or metal ledger.

Ledgerboard. The support for the second-floor joists of a balloon-frame house.

Leg. 1. A tile wall running alongside a bathtub or abutment. 2. A narrow strip of tile floor.

Legal Description. 1. The description of a piece of property such that it can be positively identified for purposes such as purchase, sale, hypothecation, title insurance, and locating it on the land. 2. The lot, block, and tract numbers of a recorded subdivision. 3. A metes and bounds description.

Length. 1. The extent from end to end of an object. 2. The longer or longest of the 2 or 3 dimensions of an object. 3. The dimension measured between the ends of a masonry unit, usually the dimension of the masonry unit which is parallel to the face or length of the wall.

Length Seam. See Side Seam.

Leno Weave. Weave in which warp yarns, arranged in pairs, are twisted around one another between picks of weft yarn.

Lens. 1. A transparent glass or plastic substance with one or both sides curved for concentrating or dispersing light rays, as in optical devices. 2. The diffuser of a lighting fixture.

Lessee. One that holds property under a lease.

Lessor. One that conveys property by lease.

Let-In Bracing. Diagonal bracing, usually 1" x 4" or 1" x 6", nailed into notches cut in the face of the studs so as to avoid an increase in the thickness of the wall.

Letter Box. A receptacle for the placement and storage of mail.

Letter of Credit. A letter written by a bank guaranteeing that drafts up to a certain amount will be honored; in some cases this will serve as a substitute for a surety bond.

Letter Sign. Sign composed of individual letters, usually of metal or plastic, mounted on a rigid background surface.

Letter Slot. An opening in a wall or door for the insertion of mail.

Level. 1. The position of a line or plane when parallel to the surface of still water. 2. A tool of wood, metal, or plastic used in testing for horizontal and vertical surfaces; see Spirit Level. 3. An instrument used in determining differences of elevation; see Engineer's Level. 4. A. roof is considered level up to a maximum slope of 1/2-inch per foot.

Level Area. A specified surface that does not have a slope in any direction exceeding 1/4 inch in one foot from the horizontal (2.083 percent gradient).

Leveler, Dock. See Dock Leveler.

Level Floor. To bring a floor in conformance with a specified grade.

Leveling. In painting, the formation of a smooth film on either a horizontal or vertical surface, independent of the method of application; a film with good leveling characteristics is usually free of brush marks or orange peel effects.

Leveling Coat. A coat or layer of plaster or stucco which brings the surface to a true level plane.

Leveling Plate. A steel plate placed in grout on top of a concrete foundation to create a level bearing surface for the lower end of a steel column.

Lever. A simple machine consisting of a rigid bar pivoted about a fulcrum (fixed point) which can be acted upon by a force in order to move a load.

Lever Handle. A horizontal handle for operating the bolt of a lock.

LH. 1. Left hand. 2. Left Hand, a door handing designation.

LH Joist. A type of long-span high strength bar joist.

LHR. Left Hand Reverse, a door handing designation.

LH Series. A standard Steel Joist Institute designation for longspan steel joists.

Liabilities. Creditor claims on the assets of a business.

Liability. Legal responsibility.

Liability Insurance. Insurance that will pay damages for which the insured becomes liable.

Library. A room or building used for the storage and use of books.

Licensed Architect. An architect duly licensed by the state to practice.

Licensed Contractor. A contractor duly licensed by the state to construct buildings.

Lien. A legal clam by a party against another party for satisfaction of a monetary claim.

Lien Claim. A claim to obtain a lien against real property because of nonpayment by the owner for labor, services, or material supplied for a work of improvement.

Lien Foreclosure Action. 1. A lawsuit to foreclose a Mechanics' Lien. 2. A lawsuit for foreclosures on property brought about to secure payment due to the holder of a lien and judgment against the real property involved.

Lien Release Notice. A notice showing that the lien amount has been satisfied and releasing the lien claim against the real property involved.

Life-Cycle Cost. A cost that takes into account both the first cost and costs of maintenance, replacement, fuel consumed, monetary inflation, and interest over the life of the object being evaluated.

Life Estate. The right to own, use, and occupy real property during one's lifetime.

Life Expectancy. The average time a material, piece of equipment, or assembly would be expected to give satisfactory service under the conditions in which it is used.

Lifeguard Chair. A raised chair, equipped with a ladder, that enables a lifeguard to view a large area.

Lift. 1. The amount of material placed at one time. 2. A layer of concrete. 3. A layer or course of paving material applied to a base or a previous layer. 4. A device to elevate loads.

Lift, Auto. See Auto Lift.

Lift, Hydraulic. See Hydraulic Lift.

Lifting. 1. In painting, the buckling of the finish coat when applied over previous coat which is not yet dry or when solvents in second coat are too strong. 2. Softening of undercoat by solvents used in coats which follow; usually caused by not allowing sufficient time for undercoat to harden before applying additional costs.

Lift-Slab Construction. A method of building multi-story sitecast concrete buildings by casting all the slabs in a stack on the ground, then lifting· them up the columns with jacks and welding them in place.

Light. 1. Daylight. 2. A lamp. 3. An appearance of brightness. 4. A pane of glass in a window. 5. Low in weight, density, or intensity.

Light, Dental. See Dental Light.

Light Duty Tile. Tile suitable for limited pedestrian traffic such as entryways in single family residences.

Light-Emitting Diode (LED). An electric component that emits light.

Lighting. An artificial supply of light or the apparatus providing it.

Lighting Contractor. The person in charge of installing an apparatus which supplies an artificial supply of light.

Lighting Fixture. An assembly having one or more lampholders, or a lampholder used in lieu of such an assembly.

Lighting Outlet. An outlet intended for the direct connection of a lampholder, a lighting fixture or a pendant cord terminating in a lampholder.

Lighting Standard. A pole supporting a lighting fixture as along streets or in a parking lot or athletic field.

Lightly Coated Electrode. A filler-metal electrode, used in arc welding, consisting of a metal wire with a light coating applied subsequent to the drawing operation, primarily for stabilizing the arc.

Lightning Arrester. A device connected to an electrical system to protect from lightning and voltage surges.

Lightning Rod. A metallic rod mounted on a high part of a structure, attached to a ground wire, to divert lightning into the earth.

Light, Pool. See Pool Light.

Light, Surgical. See Surgical Light.

Light Track. An electrified U-shaped member attached to a ceiling or wall and acting as a channel for sliding light fixtures.

Light Well. A court, open to the roof, to provide light and ventilation to rooms that face it.

Lightweight Aggregate. 1. Aggregate of low specific gravity, such as expanded or sintered clay, shale, slate, diatomaceous shale, perlite, vermiculite, or slag; natural pumice, scoria, volcanic cinders, tuff, and diatomite, sintered fly ash or industrial cinders; used to produce lightweight concrete. 2. Aggregate with a dry, loose weight of 70 pounds per cubic foot or less.

Lightweight Block. A concrete masonry unit constructed of lightweight materials and used to reduce the weight of walls.

Lightweight Concrete Firestop. Lightweight concrete used as firestop.

Lightweight Concrete. See Concrete, Lightweight.

Lignin. An amorphous polymeric substance related to cellulose that together with cellulose forms the woody cell walls of wood and the cementing material between them.

Lime. 1. Various compounds of calcium oxide and magnesium oxide, obtained by heating forms of calcium carbonate, such as shells or limestone; used in mortar and plaster; also called Caustic Lime and Quicklime. 2. A dry white powder consisting essentially of the chemical compound calcium hydroxide that is made by treating caustic lime with water. 3. Various chemical and physical forms of quicklime, hydrated lime, and hydraulic hydrated lime.

Lime Mortar. Mortar in which the cementing agent is lime.

Lime Plaster. Basecoat plaster consisting essentially of lime and an aggregate.

Lime Putty. Slaked or hydrated lime that is mixed with water to form a putty like mixture.

Lime Scale. The build-up of calcium carbonate on plumbing fixtures and in piping from the water supply.

Limestone. A sedimentary carbonate rock, composed chiefly of calcite ($CaCO_3$), but sometimes containing appreciable dolomite.

Limestone Aggregate. Granular, crushed limestone.

Limestone Lintel. Limestone member placed within masonry wall or partition to support masonry or other construction over wall or partition opening.

Limestone Panel. A limestone slab, relatively thin with respect to other dimensions, and rectangular in shape.

Limit Control. Control used to open or close electrical circuits as temperature or pressure limits are reached.

Limited Partnership. A partnership in which the management authority and liability of some of the partners is limited to their original investment; there must be at least one general partner who remains fully responsible for business liabilities and who runs the business.

Limiting Friction. The maximum value friction can have before motion ensues.

Line. 1. A length of string, thread, or rope. 2. A utility service such as a water pipe, electrical wire, or sewer pipe. 3. To cover the inside surface of an object with some material. 4. A circuit in a communication system. 5. The boundary of an area; a defining outline. 6. A straight or curved geometric element that is generated by a moving point and that has extension only along the path of the point. 7. A curve connecting all points having a specified common property. 8. A straight line is the most direct route connecting two points. 9. A string that is stretched from lead to lead to act as a guide for brick courses.

Linear Measure. Measurement along a line.

Linen Chute. An inclined plane, sloping or vertical channel, or passage down or through which soiled linen may pass.

Line Pin. Metal pin used to attach string line used for alignment of masonry units.

Liner, Flue. See Flue Liner.

Linerless Pool. Prefabricated component pool that retains water without additional waterproofing when assembled.

Liner Panel. A panel used for interior finish.

Liner, Pool. See Pool Liner.

Line Wire. Taut parallel horizontal wires installed on the outside of wood studs to act as a backing for weatherproof paper under the stucco netting of an exterior stucco wall assembly.

Lining Paper. Wallpaper without a ground (overall background color), used mostly for wall conditioning.

Lining, Carpet. See Carpet Padding.

Link, Beam. That part of a beam in an eccentrically braced frame which is designed to yield in shear and/or bending so that buckling of the bracing members is prevented.

Linoleum. A tough wearing resilient floor covering, commonly in sheet form, consisting of a burlap back coated with a preparation of linseed oil, powdered cork, and pigments.

Linoleum and Oilcoat Varnishes. Special highly flexible and elastic varnishes.

Linoleum Knife. See Hook Knife.

Linoleum Removal. The act or process of scraping away old linoleum flooring, commonly done by a specially designed apparatus which mechanically removes both linoleum and adhesive .

Linoleum Varnish. Special highly flexible and elastic varnish.

Linseed Oil. Yellowish drying Oil extracted from flax plant seed, widely used as a vehicle for lead-based paints; it is soluble in ether, benzene, and turpentine; metallic salts or driers are added to increase rate of drying; see also Bodied Linseed Oil and Boiled Linseed Oil.

Lint Strainer. A device mounted in the pump influent line to catch lint and other debris.

Lintel. A horizontal structural member, usually made of stone, wood or metal, which supports the load over an opening; a Header.

Lintel Block. A channel block.

Lintel Coarse. See String Course.

Lintel, Limestone. See Limestone Lintel.

Lintel, Masonry. See Masonry Lintel.

Lintel, Steel. Steel member placed within wall or partition to support loads over an opening.

Lip. The chain and/or stuffer left on the edge of carpet after it has been cut.

Lip Molding. A molding with a lip which overlaps the piece against which the back of the molding rests.

Lip of a Strike. The projecting part of a door lock strike on which the latch bolt rides.

Lipping. Laying brick so that the top edge of the unit is set in from the plane surface of the wall.

Liquefaction. Transformation of a granular material, the soil, from a solid state into a liquid state as a consequence of vibrations induced by an earthquake.

Liquefied Propane Gas. A compressed gas, such as propane or butane, used for fuel.

Liquid. A substance that flows freely like water and has a form that has a definite volume but no fixed shape; one of the three states of matter; compare with solid and gas.

Liquid Absorbent. Chemical in liquid form which has the property to take on or absorb other fluids.

Liquid Applied Membrane. A roofing system applied in one or more coats, suitable for intricate shapes such as cast-in-place concrete structures.

Liquid Assets. Current assets that are, or can be, readily converted to cash.

Liquidate. Wind up the affairs of a firm by ascertaining liabilities and apportioning assets; to determine by agreement or by litigation the precise amount of indebtedness, damages, or accounts.

Liquidated Damages. An amount determined by contract in advance of injury to be paid to compensate a party for an injury or damages.

Liquid Collector. A collector with a liquid as the heat transfer fluid.

Liquid Driers. Solution of driers in paint thinners.

Liquid Indicator. Device located in liquid line which provides a glass window through which liquid flow may be watched.

Liquidity. Availability of cash, cash equivalents, and readily sold assets.

Liquid Line. Tube which carries liquid refrigerant from the condenser or liquid receiver to the refrigerant control mechanism.

Liquid Nitrogen. Nitrogen in liquid form which is used as a low-temperature refrigerant in expendable or chemical refrigerating systems.

Liquid Receiver Service Valve. Two- or three-way manual valve located at the outlet of the receiver and used for installation and service purposes; also called the King Valve.

Liquid Receiver. Cylinder (container) connected to condenser outlet for storage of liquid refrigerant in a system.

Liquid-Vapor Valve Refrigerant Cylinder. Dual hand valve on refrigerant cylinders which is used to release either gas or liquid refrigerant from the cylinder.

Liquid Waste. The discharge from any fixture or appliance connected to a drainage system which does not receive fecal matter.

Liquid Wood Filler. Varnishes of low viscosity, usually containing extending pigment, for use as a first coating on open- grain woods; its purpose is to afford a non-absorbent surface for succeeding coats of varnish; it is frequently colored so as to stain and fill in one operation.

Liquor. Solution used in absorption refrigeration.

Lis Pendens. A document recorded in the county recorder's office that gives public notice of litigation involving title to or the possession of real property or real estate.

Listed. Listed means equipment or materials included in a list published by a listing agency that maintains periodic inspection on current production of listed equipment or materials and whose listing states either that the equipment or material complies with approved standards or has been tested and found suitable for use in a specified manner.

Listing. A contract with a real estate broker, for a limited period of time, to produce a buyer or lessee for real property under specified rental and conditions.

Listing. Electrical materials, devices, fixtures, fittings, equipment, appliances and accessories that are shown in a list published by the enforcing agency, or by an approved testing agency, qualified and equipped for experimental testing and maintaining adequate periodic inspection of current production of listed models, and whose listing states either that the material, device, fixture, fitting, equipment, appliance and accessory complies with nationally recognized health, fire and safety requirements, or has been tested and found safe for use in a specified manner. Compliance may be evidenced by the presence of a label or identification mark of such approved testing agency.

Listing Agency. Listing agency means an agency accepted by the Administrative Authority which is in the business of listing or labeling and which maintains a periodic inspection program on current production of listed models, and which makes available a published report of such listing in which specific information is included that the product has been tested to approved standards and found safe for use in a specified manner.

Listing Mark. An independent laboratory mark or manifest indicating the material bearing this mark may be used in test certified by that lab.

List Price. A published price, subject to possible trade discounts.

Liter. A unit of volume equal to 1000 cubic centimeters, 1000 cubic centiliters, or 1.057 quarts.

Litmus. A dye that turns red in acidic conditions and blue in alkaline conditions; litmus paper is stained with litmus to be used as a test for acids or alkalis.

Live Load. Any load that is not permanently applied to a structure; the weight of people, furnishings, machines, and goods in or on a building; the vertical load superimposed by the use and occupancy of a building; all loads on a building except dead and lateral loads.

Live Part. Live parts are those parts which are electrically connected to points of potential different from that of the earth.

Livering. Formation of curds or gelling; coagulation of varnish finishing material into a viscous, rubber-like mass; usually caused by chemical reaction of two or more non-mixing products.

Live Space. A space with reflective surfaces so that sound can be sustained through several reflections.

Live Wire. The hot wire, carrying electric current, as opposed to the neutral or ground.

Living Room. A room in a residence for the occupants' common social pursuits.

Living Standard. Standard of living.

Living Unit. A house or apartment for the use of one family.

L-Mold. A metal molding strip with an L-like section, used to edge plaster or gypsum wallboard.

Load. 1. A force that is applied to a body. 2. A weight or force acting on a structure. 3. Applied external force, such as gravity and wind. 3. Any device that consumes electrical power, such as a motor, lamp, or toaster; the load on an electrical circuit.

Load Bearing Metal Stud. Vertical load bearing, formed "C" channel of steel, component within framed wall, able to withstand structural loads imposed by wind loads, building loads, movement and deflection of structure.

Load-Bearing Partition. A vertical structural interior wall supporting an integral part of the construction above.

Load Bearing Stud. A wooden stud which supports an imposed load in addition to its own weight.

Load Bearing Wall. A wall designed and built to carry superimposed vertical and shear loads as opposed to non-load-bearing walls which carry only their own weight.

Load Bearing. Supporting a superimposed weight or force.

Load Duration. The period of continuous application of a given load, or the sum of the periods of intermittent application of the same load.

Loader. An excavating machine with a movable bucket or scoop, used to transport earth, crushed stone, or other construction materials.

Load Factor. 1. The percentage of the connected load which is likely to occur at any time. 2.The percentage of the total connected fixture-unit flow rate which is likely to occur at any point in the drainage system. 3. The percentage of the total electrical load which is likely to be in use at any time.

Load, Factored. The load, multiplied by appropriate load factors, used to proportion structural members by the strength design method.

Loading Dock Equipment. Powered or hand operated machinery to help load and unload freight.

Loading Dock. A raised platform adjacent to a loading door so that trucks may be loaded or unloaded with the truck bed at the same level as the warehouse floor.

Loading Pump. A hand pump for filling bulk materials into application tools.

Load, Service. See Service Load.

Loam. A soft, easily worked soil containing sand, silt and clay.

Loan. Money lent at interest.

Loan Fee. A fee for negotiating a loan, in addition to the interest.

Lobby. An entrance room or anteroom.

Local Access Streets. Most of the low speed controlled access roads within sectors of a city, between collector streets, serving built-up areas, parking and pedestrians unrestricted.

Local Preheating. Preheating a specific portion of a structure.

Local Stress-Relief Heat Treatment. Stress-relief heat treatment of a specific portion of a structure.

Local Ventilating Pipe. A pipe on the fixture side of the trap through which vapor or foul air is removed from a room or fixture.

Location. A particular place or position.

Lock. A mechanism for fastening a door with a bolt that requires a key of a particular shape or a combination of movements to work it.

Locker, Clothing. See Clothing Locker.

Locker Room. A room for changing clothes and storing them in lockers, especially for sporting activities.

Locknut. 1. A nut screwed hard up against another to prevent either of them from moving. 2. A nut so constructed that it locks itself when screwed up tight.

Locknut, Conduit. See Conduit Locknut.

Locknut, Grounding. See Grounding Locknut.

Lock Plug. Cylinder plug of a lock.

Lock Rail. The horizontal rail of a door intended to receive the lock case.

Lockset. A device installed in a door that has both a deadbolt and doorknob assembly.

Lock Stile. The door stile to which the lock is applied as distinguished from the hinge stile.

Lockwasher. A flat, split ring of metal or steel that when tightened with a nut is used to prevent loosening.

Locus. 1. A position or point. 2. A curve or line formed by all the points satisfying a particular equation.

Lodging House. Rented living quarters, usually in a private house rather than a hotel.

Loft. One of the upper floors of an unpartitioned warehouse building.

Log. A section of tree trunk suitable in length for sawing into commercial lumber.

Loge. 1. A small booth or compartment. 2. A box in a theater. 3. A separate forward section of a theater mezzanine or balcony.

Loggia. An opensided gallery or arcade, especially on upper floors of a building.

Logistics. The organization and details of carrying out an operation.

Log Lighter. A length of perforated pipe in a fireplace, connected to a valved gas line, used for lighting fires.

Longitudinal. 1. Running lengthwise. 2. In wood, generally, parallel to the direction of the wood fibers.

Longitudinal Bar. Any reinforcing bar placed in the long direction of the member.

Long-Oil Varnish. Varnish with a large percentage of oil to gum resin, usually more than 25 gallons of oil to 100 pounds of resin; long-oil varnish is more elastic and more durable than short-oil varnish; spar varnish is a typical example of long-oil varnish.

Long-Term Burst. The internal pressure at which a pipe or fitting will break due to a constant internal pressure held for 100,000 hours.

Long-Term Capital Gain. The gain upon sale of capital assets that have been held for over 6 months.

Long Term Creditors. Persons or companies to whom money is owed but not payable for over a year.

Long Term Debt. Mortgages and loans due in over a year.

Lookout. 1. The end of a rafter. 2. The construction which projects beyond the sides of a house to support the eaves. 3. The projecting timbers at the gables which support the rake boards.

Loom. Machine on which carpet is woven, as distinguished from other machines on which carpets may be tufted, flocked or punched.

Loomed Carpet. Carpet made on a modified upholstery loom with characteristic dense low-level loop pile, generally bonded to cellular rubber cushioning; see Woven Carpet.

Loop Pile. A Wilton or Velvet carpet woven with the yarn uncut; also called Round Wire.

Loop Vent. This is the same as a circuit vent; the only difference is that the vent loops back and connects with a soil stack or waste stack instead of a vent stack.

Loose Fill Insulation. Several types of thermal insulation in the form of fibers, granules, or other pieces that can be pumped, poured or placed by hand.

Loose Joint Hinge. A hinge having but two knuckles, the pin being fastened permanently to one knuckle, the other containing the pinhole, allowing the two parts of the hinge to be disengaged by lifting the door; such hinges are handed.

Loose Knot. A knot that is not held firmly in place by growth or position and that cannot be relied upon to remain in place.

Loose Laid. A roofing membrane not attached to the substrate.

Loose Steel Lintel. Lintel made of structural steel shape (usually angle) that is set loose and built into masonry construction.

Loss on Ignition. The loss in weight expressed as a percentage, of a sample ignited at a very high temperature (1000 ± degrees C).

Lot Line. The boundary of a parcel of land.

Lots. Parcels of land in a recorded subdivision; usually numbered, and shown on a map.

Loudness. 1. The subjective human definition of the intensity of a sound; human reaction to sound is highly dependent on the sound pressure and frequency.

Louver. 1. A construction of numerous sloping, closely spaced slats used to prevent the entry of rainwater into a ventilating opening. 2. A kind of window, generally in peaks of gables and the tops of towers, provided with horizontal slats which exclude rain and snow and allow ventilation. 3. An opening with a series of horizontal slats so arranged as to permit ventilation but to exclude rain, sunlight, or vision; see Attic Ventilators.

Louver, Door. See Door Louver.

Louver, Metal. See Metal Louver.

Louver, Penthouse. See Penthouse Louver.

Louver Vent. A construction element with equally spaced slots that allow ventilation.

Love Wave. Transverse vibration of seismic surface wave; see Seismic Wave.

Low Boiling Solvent. Solvent with a low boiling point such as acetone or methyl alcohol.

Low Consistency Plaster. A neat (unfibered) gypsum basecoat plaster especially processed so that less mixing water is required than in standard gypsum basecoat plaster to produce workability; this type plaster is particularly adapted to machine application.

Low-e Glass. Low emissivity glass.

Low Emissivity Glass. Glass that reduces energy loss by reflecting heat back into a living space.

Low-Energy Power Circuit. A circuit which is not a remote-control or signal circuit, but which has the power supply limited in accordance with the requirements of Class 2 remote control circuits. Such circuits include electric door openers and circuits used in the operation of coin-operated machines.

Lowest Responsible Bidder. The lowest bidder who is considered qualified and responsible.

Low-Iron Glass. Glass formulated with a low iron content so as to have a maximum transparency to solar energy.

Low-Lift Grouting. A method of constructing a reinforced masonry wall in which the wall is grouted in increments not higher than 4 feet.

Low Pressure Control. Cycling device connected to low-pressure side of refrigeration system.

Low Pressure Gauge. Instrument for measuring pressures in range of 0 psi to 50 psi (0 $kg/cm^2$to 3.52 kg/cm^2).

Low-Pressure Steam Boiler. A boiler furnishing hot water at pressures not more than 30 pounds per square inch gauge (2.12 kg/cm^2) or steam at pressures not more than 15 pounds per square inch gauge (1.06 kg/cm^2).

Low Relief. Bas relief.

Low Rise. A building of one or two stories and having no elevator.

Low Side. That portion of a refrigerating system which is below evaporating pressure.

Low-Side Float Valve. Refrigerant control valve operated by level of liquid refrigerant in low pressure side of system.

Low-Side Pressure Control. Device used to keep low-side evaporating pressure from dropping below certain pressure.

Low-Side Pressure. Pressure in cooling side of refrigerating cycle.

Low-Voltage Wiring. Wiring used for control, communication, and signal circuits, usually 12 or 24 volts.

Lozenge. A rhombus or diamond figure.

LP Fuel. Liquefied petroleum used as a fuel gas.

LPG. Liquefied Propane Gas.

LTC Ratio. In real estate loans, the Loan to Cost Ratio.

Lubrication. The use of oil or grease to make parts of machines move easily by reducing friction.

Lucite. Trademark, used for an acrylic resin or plastic consisting essentially of polymerized methyl methacrylate.

Lugs. Spacers, or protuberances on the sides of ceramic tiles; these devices automatically space the tile for the grout joints.

Lug Sill. A precast window or door sill for a masonry wall which is interlocked with the wall system.

Lumber. Sawed parts of a log such as boards, planks, and timber; wood members which are manufactured by sawing, resawing, passing lengthwise through standard planing machine, crosscutting to length, and matching, but without further manufacturing.

Lumber Grading. See Grade, 1.

Lumberjack. 1. A logger. 2. A Lumberman.

Lumberman. 1. One who processes, conveys, or merchandizes lumber. 2. A Lumberjack.

Lumen. 1. In wood anatomy, the cell cavity. 2. A lighting unit; the amount of light emitted through an opening of 1 square foot, located at a distance of 1 foot from a light source, which emits 1 candlepower in every direction.

Luminaire. A complete lighting unit consisting of a light source, switch, globe, reflector, housing, and wiring.

Lump Lime. Quicklime as it comes from the kiln.

Lump Sum Contract. A contract in which the amount to be paid to the contractor is agreed in advance to be a stipulated sum.

Luncheonette. A small restaurant offering light lunches.

Lunch Room. A room where workers or students eat their lunch.

Lunette. An architectural element that has the shape of a crescent or half moon, like an opening in a vault for a window or an alcove for a painting or statue.

Luster Fabric. Any cut pile fabric woven with surface yarns spun from special types of staple and chemically washed, like hand-woven Oriental fabrics, to give a bright sheen or luster.

M

Machine Direction. The direction parallel to the paper-bound edge of a sheet of gypsumboard.

Machine Room. Area where commercial and industrial refrigeration machinery, except the evaporators, is located.

Macrozones. Large zones of earthquake activity such as zones designated by the Uniform Building Code Map.

Madder. Coloring matter originally derived from the pulverized root of a plant cultivated in Europe and Asia Minor; now largely made synthetically.

Magnesium Silicate. White extender pigment which adds fluffiness to products in which it is used; provides very little opacity.

Magnet. A body having the capability of attracting iron and producing a magnetic field external to itself.

Magnetic. Having the properties of a magnet.

Magnetic Catch. A cabinet catch that uses a magnet to hold the door closed.

Magnetic Clutch. Device operated by magnetism to connect or disconnect a power drive.

Magnetic Compass. An instrument showing the direction of magnetic north and compass bearings.

Magnetic Core. Magnetic center of a magnetic field.

Magnetic Door Holder. A door holder using a magnet to hold it in an open position.

Magnetic Field. A region of variable force around magnets, magnetic materials, or current-carrying conductors.

Magnetic Gasket. Door-sealing material which keeps door tightly closed with small magnets inserted in gasket.

Magnetic Hammer. A special design hammer magnetically sensitized to hold a metal fastener during application.

Magnetic Pole. 1. Either of the two nonstationary areas in the north and south polar regions of the earth to which a magnetic compass needle will point 2. Either of the poles of a magnet.

Magnetic Starter. Automatic protective equipment using an electromagnet to operate; insures that a motor does not receive too high a current when starting up.

Magnetism. A field of force which causes a magnet to attract materials made of iron, nickel-cobalt or other ferrous material.

Magnetite. An aggregate used in heavyweight concrete, consisting primarily of ferrous metaferrite (Fe_3O_4); a black magnetic iron ore with a specific gravity of approximately 5.2 and a Mohs hardness of about 6.

Magnetron. A diode vacuum tube in which the flow of electrons is controlled by an externally applied magnetic field to generate power at microwave frequencies; the active element of a microwave oven.

Magnification Factor. An increase in lateral forces at a specific site for a specific factor or set of conditions.

Magnitude. A measure of earthquake size which describes the amount of energy released.

Mahogany Veneer. A thin layer of straight-grained medium density wood for an outer finish or decoration.

MAI. Member of the Appraisal Institute; a professional designation of a qualified real estate appraiser.

Mail Chute. An inclined or vertical channel through which mail travels from the exterior of a box or building to a container inside that box or building.

Main. The chief pipe, duct, or cable in any electrical, water, gas, sewer, vent or other utility system.

Main Breaker. 1. A switch in a main electrical service panel where the service wires attach. 2. The main electrical service protective device where the power enters a building.

Main Drain. See Main Outlet.

Main Outlet. Outlet fitting(s) at the bottom of a swimming pool, spa or hot tub through which passes water to the recirculating pump; also called Main Drain or Sump Pot.

Main Runners. The heaviest integral supporting members in a suspended ceiling; main runners are supported by hangers attached to the building structure and in turn support furring channels or rods to which lath is fastened.

Main Switch. See Service Disconnect.

Maintenance. The systematic upkeep of property or equipment.

Maintenance Equipment. Any of a variety of implements to maintain and clean areas or equipment.

Maintenance Painting. 1. Repair painting; any painting after the initial paint job. 2. All painting except that done solely for aesthetics.

Maisonette. 1. A small house. 2. An apartment on two floors.

Maisoning. The process of planning and carrying out the keying and master keying of a building; especially complex systems as in hotels and office buildings.

Majolica. 1. Earthenware with an opaque luster glaze and overglaze colored decorations. 2. Any decorated earthenware having an opaque glaze.

Make-Up Air. Fresh air brought into a building from outdoors through the ventilation system and that has not been previously circulated through the system.

Make-Up Water. Fresh water used to fill or refill a swimming pool.

Male Thread. Outside threads on a pipe or fitting.

Maleic Resins. Resins based on reaction between maleic anhydride or maleic acid with glycerine and rosins.

Malice. A conscious desire to do harm.

Malicious Prosecution. Pursuing a lawsuit without probable cause.

Mall. 1. A sheltered walk or promenade. 2. A shopping area grouped around a common pedestrian way.

Malleable. Of metals, capable of being formed into new shapes by hammering and bending.

Malleable Iron. Iron that can be hammered or bent without breaking.

Malleable Strap. A metal fastening plate that can be hammered or bent without breaking.

Mallet. A hammer with a wooden head.

Malpractice. Negligent act or omission of a professional.

Malpractice Insurance. See Errors and Omissions Insurance.

Managerial Competence. An aptitude for management and fulfilling of all the responsibilities inherent in the managerial functions such as planning, control, cooperation, and communication.

Mandrel. 1. A tapered axle inserted into a hole in a piece of work to support it during machining. 2. A metal bar used as a core around which material may be cast, molded, forged, bent, or otherwise formed. 3. The shaft and bearings on which a tool is mounted, as in a drill or circular saw.

Mandrel Test. A physical bending test for adhesion and flexibility.

Manganese. 1. A greyish-white, hard and brittle, metallic chemical element that resembles iron but is not magnetic. 2. A mineral that is contained in the clay used for brickmaking.

Manhole. A hole through which a person may go to gain access to an underground or enclosed structure.

Manhole Base. The cast iron frame into which a manhole cover fits.

Manhole, Communication. See Communication Manhole.

Manhole Cover. A heavy, round or square, steel or iron cover used to gain access to underground work through a manhole.

Manhole, Electric. See Electric Manhole.

Manhole Removal. The act or process excavating an existing manhole and removing it.

Manhole, Watertight. See Watertight Manhole.

Manifold. A pipe fitting with several lateral outlets for connecting one pipe with others.

Manifold Gauge. Chamber device constructed to hold both compound and high-pressure gauges; valves control flow of fluids through it.

Manifold, Service. See Service Manifold.

Manila Paper. See Cream Paper.

Manila Resins. Alcohol soluble natural resin obtained by tapping Agathis alba trees.

Manometer. A pressure gauge for gases and liquids.

Manor. The house of a landed estate.

Mansard Roof. A type of hip roof which has four sloping sides, each of which becomes steeper partway down.

Mansion. A large imposing residence.

Mantel. 1. A projecting shelf above the fireplace opening. 2. The entire finish around a fireplace.

Mantle. The main bulk of the earth that lies between the crust and the central core.

Manual. 1. Done with the hands. 2. A book of instructions for operating a machine or performing some task.

Manual Frost Control. Manual control used to change operation of refrigerating system to produce defrosting conditions.

Manual Proportioning Control. In an asphalt batching plant, a control system in which proportions of the aggregate and asphalt fractions are controlled by means of gates or valves which are opened and closed by manual means; the system may or may not include power assist devices in the actuation of gate and valve opening and closing.

Manufactory. A factory.

Manufacture. Making of articles as in a factory, especially on an industrial scale.

Manufactured Roof. A factory-finished roof system.

Manufactured Wall. A factory-finished wall system.

Manufacturer's Bond. A surety company's guarantee that it will stand behind a manufacturer's liability to finance roofing membrane repairs occasioned by ordinary wear within a period generally limited to 5, 10, 15 or 20 years.

Map. A two-dimensional representation of a part of the earth's surface.

Mar. To disfigure, spoil, or impair the perfection of a surface or object.

Marble. Limestone that is more or less crystallized by metamorphism, that ranges from granular to compact in texture, that is capable of taking a high polish, and that is used in architecture and sculpture.

Marble Chips. Graded aggregate of maximum hardness made from crushed marble to be thrown or blown onto a soft plaster bedding coat to produce marblecrete.

Marblecrete. A marble chip embedded finish for plaster.

Marble Flooring. A floor system using marble as its finish material.

Marble, Imported. See Imported Marble.

Marbleizing. See Faux Marble.

Marble Mosaic Tile. Tile made of small marble pieces that vary slightly in size, usually about one-half inch square.

Marble Paver. A type of igneous stone that is harder and more durable than regular bricks, used for driveways and patios for use with or without mortar.

Marble Tiles. Marble cut into tile sizes 12 inches square or less, usually 1/2 inch to 3/4 inch thick; available in several types of finishes, such as polished, honed, or split faced.

Marblite. Plaster lining of marble granules with white cement to finish concrete pools.

Marezzo. An imitation marble produced with Keene's cement to which color pigments have been added.

Marine Plywood. Plywood panels manufactured with the same glueline requirements as other exterior-type panels but with more restrictive veneer quality requirements.

Marine Varnish. Varnishes especially designed to resist long immersion in salt or fresh water and exposure to marine atmosphere.

Marked Crossing. A crosswalk or other identified path intended for pedestrian use in crossing a vehicular way.

Marketable Title. Title to land that is unencumbered by any lien that would affect its value or marketability

Market Area. The geographic area from which a real estate development expects to derive its customer base.

Market Data Approach. A real estate appraisal method by which the subject property is compared with other properties that have recently been sold or leased.

Marketing Management. Development of a marketing concept by a contractor, including such factors as advertising, copy for phone directory, letterhead for stationery, logo, and business cards.

Market Study. A forecast of future demand in a particular market area for the type of real estate project proposed for development on a specific site.

Market Value. The price of something on the open market where the buyer and seller are both knowledgeable, and under no pressure to complete the transaction.

Marl. A calcareous clay, containing approximately 30 to 65 percent calcium carbonate ($CaCO_3$), found normally in extinct fresh water basins, swamps, or bottoms of shallow lakes.

Marquee. 1.A canopy over an entrance. 2. A large tent set up for outdoor receptions.

Marquetry. Decorative work in which elaborate patterns are formed by the insertion of pieces of material, such as wood, shell, or ivory, into a wood veneer that is then applied to a surface, such as a piece of furniture.

Masking. 1. The presence of a background noise increased to a level to which a sound signal must be raised in order to be heard or distinguished. 2. When painting, protecting areas not to be painted. 3. Application of protective materials used in plastering machine applications.

Masking Power. The ability of a fired glaze to mask visually the body on which it is applied.

Masking Tape. Adhesive coated paper tape used to mask or protect parts of surface not to be finished.

Mason. 1. A skilled worker who builds by use of brick or stone set in mortar. 2. See Stonemason. 3. See Brickmason.

Masonite. Trademark. Fiberboard building panels made from steam-exploded wood fiber and binders.

Masonry. Brickwork, blockwork, and stonework.

Masonry. Construction of brick, tile, stone, or concrete block, or combination thereof, bonded together with mortar.

Masonry Accessory. Any of various components necessary for masonry construction.

Masonry Anchor. A fastening device or mechanism used in a masonry wall.

Masonry Base. The lowest course of masonry in a pier, foundation, wall or footing.

Masonry Bonded Hollow Wall. A hollow wall built of masonry units in which the inner and outer wythes of the walls are bonded together with masonry units, such as in the all-rolok and rolok- bak walls.

Masonry Brick. Bricks that are shaped and molded in various sizes and shapes.

Masonry Cement. Portland cement with dry admixtures designed to increase the workability of mortar.

Masonry Chimney. A vertical noncombustible structure with a flue or flues to remove smoke and other gases constructed of shaped or molded masonry units.

Masonry Cleaning. 1. The final removal of excess grout and concrete and mortar stains, from an exterior surface of a masonry structure. 2. In older structures, the cleaning of the surface by any of several means but commonly steam, chemical, and in some rare instances sandblasting.

Masonry Control Joint. A sawed, tooled, or formed groove in a masonry joint to regulate the location of cracking resulting from the dimensional change of different parts of the structure.

Masonry Facing Panel. A masonry structure having a decorative exterior surface.

Masonry Flashing. A thin, continuous sheet of metal, plastic, rubber or waterproof paper used to prevent the passage of water through a joint in a wall, roof, or at a chimney; the material used and the process of making watertight the roof intersections and other exposed places on the outside masonry.

Masonry Floor. Shaped or molded masonry units such as, stone, brick, tile or concrete units used for finished floor coverings.

Masonry Furring. Metal or wooden strips attached to any masonry surface on which wall boards or sheathing may be conventionally attached.

Masonry Grout. A mixture of cementitious materials and aggregates and water used to fill the hollow spaces of masonry units and cavities.

Masonry, Gypsum. See Gypsum Masonry.

Masonry Insulation. Any type of insulation for hollow masonry units.

Masonry Lintel. Masonry member placed within masonry wall or partition to support loads over an opening.

Masonry Manhole. A masonry vertical access shaft from the surface to the underground utility.

Masonry Painting. The act or process of applying paint to seal or color a masonry surface.

Masonry Paver. Shaped or molded units, composed of stone, ceramic brick or tile, concrete, or cast-in place concrete used for driveways and patios.

Masonry Plaque. A commemorative or identifying inscribed tablet made of shaped or molded units, composed of stone, brick, tile, or concrete.

Masonry Plaster. A plaster surface on masonry usually made from a cement-based mixture, commonly called stucco on newer buildings, but may be some other material on older historic structures.

Masonry Plaster. A plaster surface on masonry, usually made from a cement-based mixture, commonly called stucco on newer buildings, but may be gypsum-type material on older historic structures.

Masonry Reinforcing. Lateral steel rods, wire, or mesh placed between courses of masonry units.

Masonry Research. See Masonry Institute of America.

Masonry Restoration. The act or process of the repair of a masonry structure.

Masonry Sandblast. The act or process of abrading or cutting masonry structure surfaces using sand ejected from a nozzle at high speed by compressed air; an effective form of cleaning masonry, but used sparingly because of its inclination to remove the outer (usually glazed or polished) surface.

Masonry Tie. A reinforcing strip, bar, or wire used to link courses of masonry together or to bond them with a wood or concrete backup wall.

Masonry Unit. Any brick, tile, stone, or block used in masonry construction.

Masonry Veneer. A nonstructural tier or layer of brick or stone attached to a structural masonry or wood framed wall.

Masonry Vent. Opening in a masonry structure to provide natural ventilation.

Masonry Wall. A wall constructed of brick, stone, or concrete block.

Masonry Waterproofing. Any of a variety of materials applied to masonry structures to resist or prevent the passage of water.

Mason's Hammer. A tool used by masons and tilesetters to score and cut brick, block, stone, or tile.

Mass. 1. Quantity of matter held together so as to form one body. 2. A body of matter of indefinite shape. 3. The quantity of a body that is its measure of inertia.

Mass Concrete. 1. Concrete without reinforcing; also called plain concrete. 2. Any large volume of concrete cast in place intended to resist applied loads by virtue of mass; generally a monolithic structure incorporating a low cement factor with a high proportion of large coarse aggregate.

Mass Curing. Adiabatic curing, using sealed containers.

Mass Law. In acoustics, the law relating to the transmission loss of walls, which states that in a part of a frequency range, the magnitude of the loss is controlled entirely by the mass per unit area of the panel; also, that the transmission loss increases 1 decibel for each doubling of frequency or each doubling of the panel mass per unit area.

Master. A fully qualified, proficient, skilled, and experienced builder, plumber, electrician, plasterer, painter, or other construction trade worker.

Masterformat. The copyrighted title of a uniform indexing system for construction specifications, as created by the Construction Specifications Institute and Construction Specifications Canada, commonly called the CSI format or numbering system.

Master Key. A key that will open several locks, each keyed separately.

Masterkeying. An arrangement of door lock cylinders having individual key changes which permits them all to be operated by a single key called the master key.

Mastic. 1. High viscosity solvent-based adhesive. 2. A pasty material used as a cement, as for setting tile, or a protective coating, as for thermal insulation or waterproofing. 3. Organic tile adhesive. 4. A viscous, dough-like, adhesive substance; can be any of a large number of formulations for different purposes such as sealants, adhesives, glazing compounds, or roofing membranes; see Asphalt Mastic and Flashing Cement.

Mastic Grout. A chemical mixture of organic and inorganic ingredients forming a one part grouting composition that is used directly from the manufacturer; it is more flexible and stain resistant than cement grout.

Mat. 1. A large footing or foundation slab used to support an entire structure. 2. A grid of reinforcing bars.

Matched Lumber. Lumber that is edge-dressed and shaped to make a close tongue-and-groove joint.

Matching. Machining boards to form tongue and groove joints.

Mat, Concrete. See concrete Mat.

Material. 1. The matter from which things are made. 2. The physical things needed for construction. 3. Important, essential, and relevant.

Material Handling. The act or process of transporting materials on or to a jobsite.

Materialman. An individual or organization who supplies construction materials to a project.

Materials Used. This includes the cost of all materials used on the job; usually the largest single expense item on income and expense statement.

Materiel. The materials, equipment, and supplies used in a business or on a project.

Mat, Floor. See Floor Mat.

Mat Formwork. The temporary support for a concrete mat during the pouring or placing of concrete.

Mat Foundation. A concrete slab used as a building or equipment foundation.

Mat Glaze. A colorless or colored ceramic glaze having low gloss.

Mat Reinforcing. The placing of metal or steel bars in freshly poured concrete mat to strengthen it.

Matrix. In concrete, the material that fills the spaces between the fine and course aggregates; the cement paste.

Matter. A physical substance in general that has mass and occupies space; occurs in the state or form of solid, liquid, or gas.

Mattock. A hand implement used for digging and grubbing, with features of a pick and an adze.

Maturing Range. The time-temperature range within which a ceramic body, glaze, or other composition may be fired to yield specified properties.

Maturity Date. The date on which an obligation becomes due.

Maul. 1. A heavy hammer for driving wedges. 2. A tool like a sledge hammer with one wedge-shaped end, used to split wood.

Maulstick. A light stick with a padded leather ball at one end, held by a painter in one hand to support the other hand.

Mausoleum. A large and grand tomb.

Maximum. (Pl. maxima) The highest possible or attainable amount.

Maximum Capable Earthquake. The maximum level of earthquake ground shaking which may ever be expected at the building site within the known geological framework; this intensity may be taken as the level of earthquake ground motion that has a 10 percent probability of being exceeded in a 100-year time period.

Maximum Size Aggregate. Aggregate whose largest particle size is present in sufficient quantity to affect the physical properties of concrete; generally designated by the sieve size on which the maximum amount permitted to be retained is 5 to 10 percent by weight.

MB. Megabyte. 1,024,000 bytes.

MBH. Thousands of British Thermal Units; for example, 82 MBH = 82,000 BTU.

MBMA. Metal Building Manufacturers Association.

MC. 1. Medicine Cabinet. 2. Medium-Curing Asphalt.

MCM. Thousand circular mils, designating wire size.

MCS. Multiple Chemical Sensitivity; a condition in which a person is considered to be sensitive to a number of chemicals at very low concentrations.

MDF. Medium-Density Fiberboard.

ME. Mechanical Engineer.

Mean. 1. The middle term in a progression of values. 2. Average.

Mean Effective Pressure (MEP). Average pressure on a surface when a changing pressure condition exists.

Mean Radiant Temperature. The weighted average of all radiating surface temperatures within one line of sight.

Meat Case. A refrigerated unit for the storage and preservation of meat.

Mechanical. 1. Of or relating to machinery or tools. 2. Relating to, governed by, or in accordance the principles of mechanics. 3. Anything in the plumbing, heating, air-conditioning, or fire sprinkler trades.

Mechanical Adhesion. Adhesion between surfaces in which the adhesive holds the parts together by interlocking action.

Mechanical Application. Application of plaster mortar by mechanical means, generally pumping and spraying; distinguished from hand application with a trowel.

Mechanical Bond. 1. The physical keying of one plaster coat to another or to the plaster base. 2. Tying masonry units together with metal ties or reinforcing steel or keys.

Mechanical Cycle. Cycle which is a repetitive series of mechanical events.

Mechanical Engineer. An engineer who designs plumbing, air conditioning, and other environmental systems for buildings.

Mechanical Skimmer. See Surface Skimmer.

Mechanics Lien. A charge against real estate for the value of work or materials incorporated thereon.

Mechanical Trowel. A power machine used to smooth and compact plaster finish coats; capable of producing an extremely smooth, dense surface; consists of revolving metal or rubber blades; also called a Power Trowel.

Median. 1. The middle value in a series of values. 2. A straight line in a triangle connecting a vertex with the midpoint of the opposite side.

Median Strip. A space between two opposing lanes of traffic, as in a divided highway.

Mediation. An alternative dispute resolution method by which a mediator assists and urges the parties to find a mutually acceptable resolution to their differences; unlike an arbitrator, a mediator cannot impose a solution or secession on the parties.

Medical Equipment. Apparatus or devices used in medical applications.

Medical Gases. In a hospital, the gases that are piped throughout the building for bedside and laboratory use, such as oxygen and nitrous oxide

Medical Incinerator. A device in which medical wastes are burned.

Medical Sterilizer. An apparatus that utilizes high heat or chemicals to sterilize medical equipment.

Medical Surveillance. 1. A periodic comprehensive review of a worker's health status. 2. The required elements of an acceptable medical surveillance program for various conditions that are listed in the Occupational Safety and Health Administration standards.

Medical Utensil Washer. A mechanical washing device which sterilizes, by high heat, medical utensils for use in surgeries and other health care needs.

Medical Waste. Waste materials created by hospitals or other medical settings.

Medicine Cabinet. A cabinet, usually found in the bathroom, where articles such as toiletries and home medicines are kept.

Medieval Cities. Irregular street patterns; fortified cities were first internally subdivided to the maximum utilization of the land; often internal part of the city acted as a second fortress.

Medium-Curing (MC) Asphalt. Cutback asphalt composed of asphalt cement and a kerosene-type diluent of medium volatility.

Medium Density Fiberboard. A panel product manufactured from lignocellulosic fibers combined with a synthetic resin or other suitable binder; the panels are manufactured to a density of 31 pcf (0.50 specific gravity) to 55 pcf (0.88 specific gravity) by the application of heat and pressure by a process in which the interfiber bond is substantially created by the added binder; other materials may have been added during manufacturing to improve certain properties.

Medium Duty Tile. Tile suitable for pedestrian traffic such as entryways in multiple dwellings and lobbies.

Medium Valve. A color midway between a dark color and a light color.

Meeting Rail. See Checkrail.

Meeting Room. A room for holding meetings; a conference room.

Megahertz. One million hertz, a unit of radio frequency.

Megalopolis. 1. A large city. 2. An urban complex consisting of a metropolis and its environs.

Megawatt. 1,000,000 watts.

Megohm. A unit of measure for electrical resistance. One megohm is equal to a million ohms.

Megohmmeter. Instrument for measuring extremely high resistances, in the millions of ohms range.

MEK. Methyl Ethyl Ketone.

Melt. To change a solid into a liquid by the application of heat; or the liquid resulting from such action.

Melting Point. Temperature at atmospheric pressure at which a substance will melt.

Melting Rate In welding, the weight or length of electrode melted in a unit of time; also called Burnoff Rate.

Melt Point. The temperature at which asphalt changes from solid to liquid.

Member. A single piece in a structure, complete in itself.

Membrane. 1. A sheet material that is impervious to water or water vapor. 2. A flexible or semi-flexible roof covering or waterproofing whose primary function is the exclusion of water.

Membrane Fireproofing A lath and plaster membrane having among its functions that of providing a barrier to fire and intense heat.

Membrane, Neoprene. See Neoprene Membrane.

Membrane Roof. A roof structure with a covering of a sheet material that is impervious to water or water vapor; commonly a single sheet of material.

Membrane Waterproofing. A membrane, usually made of built-up roofing or sheet material, to provide a positive waterproof floor over the substrate, which is to receive a tile installation using a wire reinforced mortar bed.

Meniscus. The upper curved surface of a liquid in a tube, concave when the liquid wets the tube, convex when it does not.

MEP. Mean Effective Pressure.

Mercalli Scale. System used to determine the location of the epicenter of an earthquake on the basis of defining zones of intensity by observations of damage by persons experiencing the earthquake.

Mercoid Bulb. Electrical circuit switch which uses a small quantity of mercury in a sealed glass tube to make or break electrical contact with terminals within the tube.

Mercury. A silvery-white heavy liquid metallic element used in barometers, thermometers, amalgams, and electrical switches; the only metal that is liquid at ordinary room temperature; chemical symbol Hg.

Mercury Fixture. A light fixture that has an electric discharge lamp that produces a blue-white light by creating an arc in mercury vapor enclosed in a tube or globe.

Mercury Switch. An electrical switch that has mercury enclosed in a vial to make a silent contact.

Meridian Lines. Part of a governmental land surveying grid system covering the country, with base lines running east and west and meridians running north and south; townships are located with reference to a specific base line and meridian; for example, T3N, R12W, MDBM is read Township 3 North, Range 12 West, Mount Diablo Base Line and Meridian.

Mesh. The square opening of a sieve.

Mesh, Reinforcement. See Welded Wire Fanbric.

Mesh, Slab. See Slab Mesh.

Mesh Tape. See Joint Reinforcing Mesh.

Mesh Tie. A wire used to hold sheets of mesh together so they will not move or spread apart when concrete is poured over the mesh.

Mesh Wire. A series of longitudinal and transverse wires arranged at right angles to each other in sheets or rolls, used to reinforce mortar and concrete; welded-wire fabric.

Message Tube. A cylindrical tube typically installed in the walls of a building or structure, operated by the suction of air, to transport messages placed in containers.

Met. Term applied to the heat release from a human at rest. It equals 18.4 Btu/sq. ft./hr. or 50 kcal/m^2/hr.

Metal. 1. Any of various fusible, ductile and typically lustrous chemical elements that can conduct heat and electricity. 2. Any alloy.

Metal Anchor. A bolt or fastener made of metal.

Metal Beam Anchor. Formed steel component used to anchor one end of a beam to another beam, girder, or column and prevent displacement of the beam under lateral or uplift loads.

Metal Bridging. Diagonal or longitudinal metal members used to keep horizontal wood members properly spaced, in lateral position, vertically plumb, and to distribute load.

Metal Building. A building or structure constructed of a structural steel frame covered by metal roof and wall panels; commonly prefabricated in a factory and assembled at the site.

Metal Chimney. Vertical metal structure with one or more flues to carry smoke and other gases or combustion into atmosphere.

Metal Clad Cable. Electrical conduit made of a flexible steel jacket wrapped around insulated wires.

Metal-Clad Cable. A fabricated assembly of insulated conductors in a flexible metallic enclosure.

Metal Clad Door. A flush wooden door covered in sheet metal.

Metal Clip Angle. Short length of metal angle used to attach two components.

Metal Cripple Stud. Less than full height metal stud, such as under or above an opening.

Metal Decking. Light-gauge, corrugated sheets used in the construction of roofs or floors.

Metal Door Buck Anchor. Formed steel component used to anchor door bucks or jambs to concrete or masonry construction.

Metal Ductwork. The light sheet metal material out of which the ducts of an HVAC system are manufactured. METAL ELECTRODE. A filler or non-filler metal electrode used in arc welding consisting of a metal wire with or without a covering or coating.

Metal-Electrode-Arc Welding. An arc-welding process wherein metal electrodes are used.

Metal Fabrication. The building, construction or manufacture of metal structures or metal devices.

Metal Floor Track. Horizontal channel shaped steel member located at bottom of framed wall to receive metal studs.

Metal Framing Anchor. Fabricated metal devices used to transfer structural loads from a wooden structural member to another or other material member or supporting structure.

Metal Framing. The construction of a building or structure by using steel; the construction of frame houses and partitions by using light gauge metal studs and members.

Metal Furring. A length of metal channel attached to a masonry or concrete wall to permit the attachment of finish materials to the wall.

Metal Girder Anchor. Formed steel component used to anchor end of girder to another girder, beam, or column and prevent displacement of girder under lateral or uplift loads.

Metal Grating. Metal screening made from sets of parallel bars placed at right angles to each other to allow water to drain through, while protecting persons and vehicles from the drain opening.

Metal Halide Lamp. A lamp that uses an electric-discharge to produce light from a metal vapor such as sodium or mercury.

Metal Joist Anchor. Formed steel component used to anchor end of joists and to wall plate or beam and prevent displacement of joist under lateral or uplift loads.

Metal Joist Bridging. Diagonal or longitudinal metal members used to keep metal joists properly spaced, in lateral position, vertically plumb, and to distribute load.

Metal Joist Hanger. Formed steel component used to support end of load bearing joists and transmit loads to another joist or beam.

Metal Joist. Horizontal cold formed metal framing member of floor, ceiling or flat roof to transmit loads to bearing points; often refers to a Bar Joist.

Metal Lath. See Expanded Metal Lath.

Metallic Color Tile. Tile that has been coated with metal flakes to reflect light.

Metallic Grout. Grout that has been coated with metal flakes that reflect light.

Metallic Soap. A compound of metal and organic acid; used as driers, fungicides, suspending agents, and flatting agents.

Metal Lintel. A horizontal metal member spanning and carrying the load above an opening.

Metallize. 1. To coat, treat or combine with a metal. 2. Coat with a thin layer of metal.

Metal Louver. A framed opening in a wall or door with fixed or movable flaps, manufactured from metal.

Metallurgy. The science concerned with the production, purification, and properties of metals and their application.

Metal Pan Stair. A stair assembly constructed to hold precast or cast-in place concrete, masonry, or stone in sheet metal pans at the treads and landings.

Metal Primer. First coating applied in finishing metal.

Metal Quarry Tile Rack. Racks that are available in many patterns and made to order for special patterns, used to maintain uniform joint widths between quarry tiles.

Metal Rafter. Horizontal cold formed metal framing member for floor, ceiling or roof to transmit loads to bearing points.

Metal Railing Bracket. Metal wall bracket to support a railing.

Metal Railing. A guard or open fence with rails and posts made of metal.

Metal Roof Removal. The act or process of tearing off and carrying away an old metal roof.

Metal Shear Plate. Metal plates used to transfer shear loads between timber members.

Metal Shingle. 1. A roof covering unit manufactured from metal and applied in an overlapping pattern. 2. Metal material used as an exterior wall finish over sheathing.

Metal Sleeper Clip. Metal clip used to secure wood sleepers to a concrete floor.

Metal Stair Pan. A stair assembly constructed to hold precast or cast-in place concrete masonry or stone in metal sheet pans at the treads.

Metal Stair. A single or series of metal steps with framework connected at landings.

Metal Stud Bracing, Horizontal. Horizontal element used to provide stiffness and to prevent buckling or rotation of metal studs.

Metal Stud Bracing. Element used to prevent buckling or rotation of metal studs.

Metal Stud. Vertical formed steel channel within a framed wall.

Metal Tie Strap. Formed steel component used to tie one wood framing member to another.

Metal Timber Connector. Fabricated metal devices used to transfer structural loads from a timber member to another timber or other material member or supporting structure.

Metal Toilet Partition. A prefinished, manufactured dividing wall in a toilet room.

Metal Toothed Ring. Metal rings with toothed edge to embed in wood to resist shear.

Metal Track. Horizontal channel-shaped steel member located at top or bottom to receive metal studs.

Metal Trim. See Casing and Bead, 4.

Meter. 1. An instrument for measuring consumption of gas, electrical current, or water. 2. A metric unit that equals 100 centimeters or 1000 millimeters and is equivalent to 39.37 inches.

Meter Center. The room or area where meters for measuring the usage of water, electricity and/or gas are located.

Meter Center, Indoor. See Indoor Meter Center.

Meter, Electric. See Electric Meter.

Meter, Gas. A mechanical measuring and recording device of the volume of gas passing a given point.

Meter, Gas. See Gas Meter.

Metering. The mechanical process of measuring the usage of water, electricity or gas.

Meter Socket. A receptacle with electric contacts in which an electric meter is plugged.

Meter Stop. A valve used on a water main between the street and a water meter; it permits installation or removal of the meter.

Meter, Water. See Water Meter.

Metes and Bounds. A type of legal description of land where each segment of the boundary line is described by length and bearing.

Methane. A colorless, odorless, highly inflammable gaseous hydrocarbon, sometimes known as marsh gas or firedamp, produced by decomposition of organic matter in marshes and mines or the carbonization of coal, used as a fuel and raw material in chemical synthesis

Method of Joints. Mathematical procedure for determining the forces in the members of a truss or frame.

Methods of Sections. Mathematical procedure for determining the forces in the members of a truss or frame.

Methyl Acetate. An inflammable fragrant liquid used as a lacquer solvent.

Methyl Alcohol. Poisonous alcohol obtained by destructive distillation of wood.

Methyl Ethyl Ketone (MEK). A flammable liquid cleaning solvent.

Methyl Formate. Low pressure refrigerant.

Methyl Isobutyl Ketone (MIBK). A strong flammable organic solvent.

Metopes. See Tryglyphs and Metopes.

Metric System. A standard measuring system based on the meter, decimally subdivided into centimeters and millimeters.

Mexican Paver Tile. Handmade terra cotta-like tile, used mainly for floors; they vary in color, texture and appearance, from tile to tile and within each tile; available in squares up to 12 inches, hexagon, octagon, elongated hexagon, fleur de lis and other shapes; coated with various types of sealers because of their soft absorptive characteristics; the coatings provide a wearing surface on the pavers which would otherwise powder away under wear.

Mezzanine. A low-ceilinged story between two main stories of a building; an intermediate story that projects in the form of a balcony.

Mho. The practical unit of electrical conductance equal to the reciprocal of the ohm.

MI. Malleable Iron.

MIA. 1. Marble Institute of America, 33505 State Street, Farmington, MI 48335, (810) 476-5558. 2. Masonry Institute of America, formerly Masonry Research.

MIBK. Methyl Isobutyl Ketone.

MI Cable. Mineral insulated sheathed cable.

Mica. Any of a group of silicate minerals with a layered structure.

Mica Pigment. Extender pigment made from silicates of aluminum and potassium which are split into very thin plates or sheets; used as reinforcing pigment since it tends to reduce checking and cracking.

Micro. One millionth part of unit specified.

Microbiologicals. Agents derived from or that are living organisms (for example, viruses, bacteria, fungi, and mammal and bird antigens) that can be inhaled and can cause many types of health effects including allergic reactions, respiratory disorders, hypersensitivity diseases, and infectious diseases.

Microfarad. Unit of condenser electrical capacity equal to 1/ 1,000,000 farad.

Microliter. One millionth of a liter.

Micrometer. 1. A gauge for accurately measuring small distances or thicknesses. 2. A unit of length equal to one millionth of a meter; also called a Micron.

Micron. A micrometer, one millionth of a meter.

Micron Gauge. Instrument for measuring vacuums very close to a perfect vacuum.

Microphone. An instrument that turns sound waves into electric current for the purposes of transmitting or recording music or voice.

Microregionalization. Breaking up of macrozones into much smaller zones of specific earthquake intensity and activity.

Microscope. An optical device found in a laboratory which has a surface for placing materials to be examined, and a lens which magnifies those materials.

Microvolt. One millionth of a volt.

Microwatt. One millionth of a watt.

Microwave Oven. An oven in which food is cooked by the heat produced as a result of short electromagnetic wave penetration of the food; the active element is the magnetron.

Middle Strip. The zone of a two-way concrete slab that lies midway between columns.

Midget Trowel. A small hand trowel used for pointing and small confined areas.

Mil. A unit of length or thickness equal to 1/1000th of an inch.

Mildew. 1. A discoloration caused by fungi. 2. A parasitic fungus growth occurring in insufficiently vented and damp surface areas.

Mildewcide. Substance poisonous to mildew; prevents or retards growth of mildew.

Mild Steel. Steel containing less than three-tenths of one percent carbon, not used as structural steel because of its low strength.

Mile. A measure of distance equal to 5280 feet.

Milieu. Environment or setting.

Military Crest. A ridge that interrupts the view between a valley and a hilltop.

Mill. 1. A building used for grinding, machining, manufacturing, or woodworking. 2. A machine for grinding or crushing.

Milli. Combining form denoting one thousandth (1/1,000); for example, millivolt means one thousandth of a volt.

Milliammeter. An instrument that measures very small amounts of electrical current.

Milliampere. One thousandth of an ampere.

Millibar. One thousandth of a bar, the metric unit of atmospheric pressure equivalent to 100 pascals.

Millimeter. A metric unit of distance measurement equal to 1/10 of a centimeter or 1/1000 of a meter.

Millimicron. 1. A unit of length equal to one thousandth of a micrometer. 2. A nanometer.

Millisecond. One thousandth of a second.

Mill-Mixed Plaster. See Gypsum Ready Mixed Plaster.

Mill Scale. The oxide layer formed during the hot rolling of metals, such as that formed on hot-rolled reinforcing bars.

Mill White. White paint used to augment illumination on interior wall surface of industrial plants, office and school buildings; the vehicle is usually of the varnish type.

Millwork. Generally, all wood materials manufactured in millwork plants and planing mills, including such items as inside and outside doors, window and door frames, blinds, mantels, panel work, stairways, moldings and interior trim, but not including flooring, ceiling, or siding.

Milori Blue. An iron blue pigment.

Mineral Aggregate. Aggregate consisting essentially of inorganic nonmetallic materials.

Mineral Black. A natural black pigment based on graphite.

Mineral Dust. The portion of the fine aggregate passing the 0.075mm (No. 200) sieve.

Mineral Fiber Felt. A building felt with mineral wool as its principle component.

Mineral Fiber Tile. A pre-formed ceiling tile that is composed of mineral fiber and a binder; has good thermal properties and acoustics.

Mineral Filler. A finely divided mineral product at least 70 percent of which will pass a 0.075mm (No. 200) sieve; pulverized limestone is the most commonly manufactured filler, although other stone dust, hydrated lime, portland cement, and certain natural deposits of finely divided mineral matter are also used.

Mineral Granules. Opaque, natural or synthetically colored aggregate commonly used to surface cap sheets, granule-surfaced sheets, and roofing shingles.

Mineral Insulated Cable. An electric cable insulated with a mineral sheathing.

Mineral Oil. An oil obtained from mineral sources as opposed to vegetable; usually petroleum or one of its distillate products.

Mineral Spirits. A clear distillate of petroleum, a solvent for asphaltic coatings.

Mineral Surfaced Roofings. Built-up roofing materials whose top ply consists of a granule-surfaced sheet; this is an asphalt saturated felt that is coated on one or both sides and is surfaced on the weather exposed side with mineral granules.

Mineral Surfaced. Factory made roofing cap sheet or shingles covered with mineral granules.

Mineral Wool Insulation. Insulation manufactured from a lightweight vitreous fibrous material.

Mineral Wool. Any of various lightweight fibrous materials used in heat and sound insulation.

Minimal. Very minute or slight.

Minimall. A small regional shopping center.

Minimum. (Pl. minima) The least possible or attainable amount.

Miniscule. Very small.

Minor Changes in the Work. Changes ordered by the architect that are within the intent of the contract documents and do not involve a change in the contract sum or the contract time.

Minus Pressure. See Negative Pressure, 2.

Minute. 1. A unit of measurement of angles, equal to 1/60 of a degree. 2. A unit of time, equal to 1/60 of an hour 3. Extremely small or tiny.

Mirror. Polished surface that forms images by reflection of light rays, usually fabricated of glass with silver coating on reverse side.

Mirror Frame. Shop fabricated or field applied perimeter trim to contain mirror and protect and conceal edges of glass, usually constructed of metal or wood.

Mirror, Plate. Plate Mirror.

Miscellaneous ACM. Interior asbestos-containing building material on structural components, structural members or fixtures, such as floor and ceiling tiles; does not include surfacing material or thermal system insulation.

Miscellaneous Taxes Payable. Estimated taxes incurred during an accounting period and owed to local and State agencies.

Miscibility. Capability of being mixed or made homogeneous.

Misconduct. Wrongful conduct.

Misses. In painting, holidays; skips; voids.

Mistake. A legal doctrine under which formation of a contract may be prevented if a party entered into the contract under a material mistake of fact.

Mist-Coat. Thin tack coat; thin adhesive coat.

Miter. The junction of two pieces of carpet, wood, or other material at an angle; usually 45°, to form a right angle, but may be any combination of angles.

Miter Box. A device for guiding a handsaw in the correct angle for making a miter joint in wood, plastic, or metal.

Miter Saw. A fine toothed saw, usually with a stiffened back, for cutting miters.

Mix. The act or process of mixing; also mixture of materials, such as mortar or concrete.

Mix Design. Devising the proportioning of water, portland cement, fine and coarse aggregates, and admixtures in a concrete mix.

Mixer. A machine used for blending the constituents of concrete, grout, mortar, cement paste, or other mixtures.

Mixer, Colloidal. A mixer designed to produce colloidal grout.

Mixer Efficiency. In concrete making, the adequacy of a mixer in rendering a homogeneous product within a stated period; homogeneity is determinable by testing for relative differences in physical properties of samples extracted from different portions of a freshly mixed batch.

Mixer, Horizontal Shaft. See Horizontal Shaft Mixer.

Mixer, Kitchen. A kitchen utensil which blends or combines food ingredients.

Mixer, Non-Tilting. A horizontally rotating drum mixer that charges, mixes, and discharges without tilting.

Mixer, Pan. See Mixer, Vertical Shaft.

Mixer, Plant. An operating installation of equipment including batchers and mixers as required for batching or for batching and mixing concrete materials; also called mixing plant when equipment is included.

Mixer, Tilting. A rotating drum mixer that discharges by tilting the drum about a fixed or movable horizontal axis at right angles to the drum axis; the drum axis may be horizontal or inclined while charging and mixing.

Mixer, Vertical Shaft. A cylindrical or annular mixing compartment having an essentially level floor and containing one or more vertical rotating shafts to which blades or paddles are attached; the mixing compartment may be stationary or rotate about a vertical axis.

Mixing Chamber. That part of a gas-welding or oxygen-cutting torch wherein the gases are mixed prior to combustion.

Mixing Cycle. The time taken for a complete cycle in a batch mixer.

Mixing Faucet. Separate faucets having a common spout providing control of the water temperature.

Mixing, Open-Top. A truck-mounted mixer consisting of a trough or a segment of a cylindrical mixing compartment within which paddles or blades rotate about the horizontal axis of the trough.

Mixing Speed. Rotation rate of a mixer drum or of the paddles in an open-top, pan, or trough mixer, when mixing a batch; expressed in revolutions per minute, or in peripheral feet per minute of a point on the circumference at maximum diameter.

Mixing Time. The period during which the constituents of a batch of concrete are mixed by a mixer; for a stationary mixer, time is given in minutes from the completion of mixer charging until the beginning of discharge.

Mixing Valve. A valve that permits mixing of liquids or a liquid and a gas.

Mixing Water. The water in freshly mixed sand-cement grout, mortar, or concrete, exclusive of any previously absorbed by the aggregate.

Mixture. The assembled, blended, co-mingled ingredients of mortar, concrete, or the like; or the proportions for their assembly.

Mnemonic. 1. Tending to assist memory. 2. Markings on fabricated parts to assist in assembling or installing in the right place.

Mobility. The degree to which a material flows.

Mobilization. 1. The act of putting into movement or circulation. 2. The assembly and movement of equipment to a jobsite.

Modacrylics. Modified acrylics; see Acrylics.

Modal Analysis. Determination of earthquake design forces based upon the theoretical response of a structure in its several modes of vibration to excitation.

Mode. 1. A manner of style or fashion. 2. A way of doing something. 3. The most frequent value in a set of data. 4. The slope of the vibration curve

Model. The original from which a mould or copy is made.

Modem. A device that connects a computer to the telephone system.

Modification. An agreed change to the terms of a contract.

Modified Bitumen. Bituminous membranes modified with styrene butadiene or atactic polypropylene to improve flexibility, elasticity, cohesive strength, resistance to flow and toughness.

Modified Mercalli. Modification of the original Mercalli Scale to represent construction materials and methods in the United States compared to the European construction methods and materials of the original Mercalli Scale.

Modular. Of or consisting of modules.

Modular Dimensional Standards. Dimensional standards approved by the American Standards Association for all building material and equipment, based upon a common unit of measure of four inches, known as the module; this module is used as a basis for the grid which is essential for dimensional coordination of two or more different materials.

Modular Masonry Unit. One whose nominal dimensions are based on the four inch module.

Modular Ratio. The ratio of modulus of elasticity of steel (Es) to that of concrete (Ec) usually denoted by the symbol n.

Modulating. Type of device or control which tends to adjust by increments (minute changes) rather than by either full on or full off operation.

Modulating Refrigeration System. Refrigerating system of variable capacity.

Module. A standardized part or independent unit used in construction.

Modulus of Deformation. A concept of modulus of elasticity expressed as a function of two time variables; strain in loaded concrete as a function of the age at which the load is initially applied and of the length of time the load is sustained.

Modulus of Elasticity. The ratio of the unit stress in a material to the corresponding unit strain; the ratio of normal stress to corresponding strain for tensile or compressive stresses below the proportional limit of the material; referred to as elastic modulus of elasticity, Young's modulus, and Young's modulus of elasticity; denoted by the symbol E.

Modulus of Rigidity. The ratio of unit shearing stress to the corresponding unit shearing strain; referred to as shear modulus and modulus of elasticity in shear; denoted by the symbol G.

Modulus of Rupture. A measure of the ultimate load-carrying capacity of a beam and sometimes referred to as rupture modulus, or rupture strength; it is calculated for apparent tensile stress in the extreme fiber of a transverse test specimen under the load which produces rupture.

Modulus of Subgrade Reaction. Ratio of load per unit area of horizontal surface (of a mass of soil) to corresponding settlement of the surface.

Mogul Base. A screw-in style base for an incandescent lamp of generally 300 watts or more.

Moist Air Curing. Curing with moist air at atmospheric pressure and a temperature of about 70° F.

Moist Cure. See Cure, 4.

Moist Room. A room in which the atmosphere is maintained at a selected temperature (usually 23.0° C ±1.7° C or 73.4° F ±3.0° F) and a relative humidity of at least 98 percent, for the purpose of curing and storing cementitious test specimens; the facilities must be sufficient to maintain free moisture continuously on the exterior of test specimen.

Moisture. Finely divided particles of water; vapor.

Moisture, Absorbed. See Absorbed Moisture.

Moisture Barrier. A membrane used to prevent the migration of liquid water through a floor or wall.

Moisture Content. 1. The amount of water in a substance; in wood, it is usually expressed as a percentage of the weight of the ovendry wood.

Moisture Content of Wood. The amount of water contained in the wood, usually expressed as a percentage of the weight of ovendry wood.

Moisture Expansion. An increase in dimension or bulk volume of a ceramic article caused by reaction with water or water vapor; this reaction may occur in time at atmospheric temperature and pressure, but is expedited by exposure of the article to water or water vapor at elevated temperatures and pressures.

Moisture Indicator. Instrument used to measure moisture content of a refrigerant.

Moisture Meter. An instrument that measures electrical resistance in a material and equates it to the amount of ambient moisture content; often used by painters to judge the suitability of plaster, masonry, or other surface for the application of paint.

Moisture Movement. The movement of moisture through a porous medium; the effects of such movement on efflorescence and volume change in hardened cement paste, mortar, concrete, or rock.

Moisture Protection. The act or process of retarding the seepage of moisture.

Moisture Vapor Transmission (MVT). Moisture vapor transmission rate through a membrane.

Mold. A divider containing a cavity into which neat cement, mortar, or concrete test specimens are cast; a form used in the fabrication of precast mortar or concrete units (e.g., masonry units); also spelled mould.

Mold, Casting. See Casting Mold.

Molded Plywood. Plywood made to some desired shape other than perfectly flat; often this shaping is done at the time the layers are glued together; two ways of molding plywood are by applying fluid pressure and with curved forms.

Molding. A strip of wood, metal or plastic trim, used to conceal joints or provide decoration; also spelled moulding.

Molding, Base. See Base Molding.

Molding, Bed. See Bed Molding.

Molding, Lip. See Lip Molding.

Molding, Picture. See Picture Molding.

Molding Plaster. A fast-setting gypsum plaster used for the manufacture of cast ornaments.

Molding, Rake. See Rake Molding.

Mold Oil. A mineral oil that is applied to the interior surface of a clean mold, before casting concrete or mortar therein, to facilitate removal of the mold after the concrete or mortar has hardened.

Mold, Running. See Running Mold.

Mole Run. A meandering ridge in a roof membrane not associated with insulation or deck joints.

Molecule. The smallest fundamental unit of a chemical compound that can take part in a chemical reaction; a group of atoms that are linked together.

Molliers Diagram. Graph of refrigerant pressure, heat, and temperature properties.

Molybdenum. A metallic element that resembles chromium and tungsten in many properties, used in strengthening and hardening steel.

Moment. A measure of the tendency to produce rotational motion, equal to the result of multiplying a the magnitude of a force) by its perpendicular distance from a particular axis or point; Bending Moment.

Moment Connection. A connection between two structural members that is resistant to rotation between the members, as differentiated from a pin connection, which allows rotation.

Moment Diagram. A graphical method of representing the value of the bending moment at any point along a beam.

Moment Distribution. A method of structural analysis for continuous beams and rigid frames whereby successive converging corrections are made to an assumed set of moments until the desired precision is obtained; also known as the Hardy Cross method.

Moment of Inertia. The summation of the products obtained by multiplying each individual unit of area by the square of its distance to an axis.

Moment, Negative. See Negative Moment.

Moment, Positive. See Positive Moment

Moment Resisting Frame. A structural frame composed of rigid joints in which the members and joints resist both vertical and horizontal forces.

Momentum. 1. The quantity of motion of a moving body, measured as a product of its mass and velocity. The impetus gained by movement.

Monastery. A house for persons under religious vows.

Monel Metal. An alloy of nickel with copper, aluminum, and iron that is resistant to corrosive liquids.

Monitor. A structure raised above the general roof level to provide vertical windows for light and air.

Monitor Cabinet. A cabinet whose doors have louvered panels to allow for ventilation, light, or finish design.

Monkey Wrench. An adjustable wrench.

Mono Pitch Truss. A truss which would develop a shed type roof.

Monochlorodifluoromethane. Popular low temperature refrigerant, Freon 12 or R-22, with a boiling point of –41° F. at atmospheric pressure; cylinder color code is green.

Monochromatic. Having or consisting of one color or hue.

Monochromatic Harmony. Color harmony formed by using shades and tints of a single color.

Monochrome Decoration. A single color decoration.

Monocoque. A type of construction in which the outer skin carries all or a major part of the stresses.

Monocottura. A single-fired tile manufactured by a process which allows the simultaneous firing of the clay with the glaze producing a finished tile with a single firing.

Monolith. A body of plain or reinforced concrete cast or erected as a single integral mass or structure.

Monolithic Concrete. Concrete cast with no joints other than construction joints or as one piece; generally, the term is used on larger structures.

Monolithic Construction. Constructed as one piece.

Monolithic Tank. One-piece shell structure.

Monolithic Terrazzo. The application of a 5/8 inch (15 mm) terrazzo topping directly to a specially prepared concrete substrata, eliminating an underbed.

Monolithic Topping. On concrete flatwork; a higher quality, more serviceable topping course placed promptly after the base course has lost all slump and bleeding water.

Monomer. 1. A chemical compound that can undergo polymerization. 2. An organic liquid, of relatively low molecular weight, that creates a solid polymer by reacting with itself or other compounds of low molecular weight or both.

Monomolecular. Composed of single molecules; substances that are one molecule thick.

Monomolecular Film. Films that are one molecule thick; used over bleeding water at the surface of freshly placed concrete or mortar as a curing agent, as it is a means of reducing the rate of evaporation.

Month-to Month Tenancy. A lease of real property, written or oral, that provides for rent to be paid monthly, for no stated term, and would require 30 days notice by either party to terminate.

Montmorillonite. See Montmorillonoid.

Montmorillonoid. A group of clay minerals, including montmorillonite, characterized by a sheet-like internal molecular structure; consisting of extremely finely-divided hydrous aluminum or magnesium silicates that swell on wetting, and shrink on drying.

Monument. 1. A commemorative structure or building. 2, A permanent surveyor's benchmark.

Monumental. 1. Of or relating to a monument. 2. Extremely great; massive; outstanding.

Mop and Flop. An application procedure in which roofing elements, such as insulation boards, felt plies, and cap sheets, are initially placed upside down adjacent to their ultimate locations, are coated with adhesive, and are then turned over and applied to the substrate.

Mop Basin. A floor set service sink; also called a Mop Receptor.

Mopping. 1. A layer of hot pitch or asphalt between plies of roofing felt. 2. Swabbing, as with roofing asphalt.

Mop Receptor. See Mop Basin.

Mop Sink. A deep well plumbing fixture with a faucet and a drain used for collecting and dispensing water for mopping and other janitorial purposes.

Moratorium. A legally authorized or prohibited suspension of contractual obligations, such as debt repayment.

Moresque. Multicolored yarn made by twisting together two or more strands of different shades or colors.

Morphology. The study of the forms of things.

Mortar. 1. A mixture of cement, sand and water; when used in masonry construction, the mixture may contain masonry cement, or standard portland cement with lime or other admixtures which may produce greater degrees of plasticity and/or durability. 2. A plaster mix. 3. The cement, sand, and water mixture that fills the interstices between gravel aggregates in a concrete mix.

Mortar Board. A square shaped board, about 3 feet square, that is used to receive the mortar on a scaffold, for the use of the mason; also called Mud Board.

Mortar Bond. The adhesion of mortar to masonry units.

Mortar Box. The box used to mix mortar.

Mortar Hoe. The mortar hoe is used for hand-mixing mortar, usually with a perforated blade and a handle about 66 inches in length.

Mortar Mix. The amount of each material specified, portland cement, fine aggregate, admixture, and water.

Mortar Mixer. A mechanical device for the mixing of mortar; most are driven by gasoline combustion engines; electrically driven mixers are used when small batches of mortar are needed.

Mortar Pumping Machine. The mortar pumping machine is used with the mortar mixer; mixed mortar is poured into the hopper, and a pneumatic gun forces the mortar through a hose; the mortar can be delivered through the hose to masons and tilesetters working high above street level; asbestos fines are added to the mortar as a bonding mechanism so that the mortar in the hose will not separate.

Mortgage. A lien against real estate that secures payment of a debt.

Mortgagee. The lender under a mortgage.

Mortgagor. The borrower under a mortgage.

Mortise Lockset. A lockset assembly with the mechanism installed in a mortised pocket within the door and frame.

Mortise. The hole which is to receive a tenon, or any hole cut into or through a piece by a chisel; generally of rectangular shape.

Mortise-and-Tenon. A joint in which a tongue-like protrusion (tenon) on the end of one piece is tightly fitted into a rectangular slot (mortise) in the side of the other piece.

Mortuary. A building in which human remains are kept until cremation or burial.

Mosaic. 1. Tile with small inlaid pieces of porcelain or natural clay materials to form decorative patterns. 2. Small tile or bits of tile, stone, or glass used to form a surface design or an intricate pattern.

Moss, Peat. See Peat Moss.

Motel. A roadside hotel providing lodging and automobile parking.

Motion. Application to a judge or arbitrator for an order or ruling.

Motor. 1. A machine, electrical or internal combustion, supplying motive power for a vehicle or some other device. 2. Rotating machine that transforms fluid or electric energy into a mechanical motion.

Motor Burnout. Condition in which the insulation of an electric motor has deteriorated by overheating.

Motor, Capacitor. See Capacitor Motor.

Motor Control Switches. Any type of switching device which is used in a motor.

Motor Control. Device to start or stop a motor or hermetic motor compressor at certain temperature or pressure conditions.

Motor, Four Pole. See Four Pole Motor.

Motor-Generator Set. A portable gasoline motor combined with an electrical generator.

Motor Grader. A self-propelled machine for earth grading.

Motor Starter. High-capacity electric switches usually operated by electromagnets.

Motor Stator. Stationary part of an electric motor.

Motor, Two-Pole. See Two-Pole Motor.

Mottling. Speckling; a non-uniform paint color.

Mould. See Mold

Moulding. See Molding.

Mounted Tile. Tile assembled into units or sheets by suitable material to facilitate handling and installation; tile may be face-mounted, back-mounted or edge-mounted; face-mounted tile assemblies may have paper or other suitable material applied to the face of each tile, usually by water soluble adhesives so that it can be easily removed after installation but prior to grouting of the joints; back-mounted tile assemblies may have perforated paper, fiber mesh, resin or other suitable material bonded to the back and/or edges of each tile which becomes an integral part of the tile installation.

Movable Louver. An element for ventilation with equally spaced slats of wood or metal that can be opened or closed.

Movable Partition. A dividing wall that can be moved and arranged to form different walled spaces.

Moving, Shrub. See Shrub Moving.

Moving, Tree. See Tree Moving.

MPT. Male pipe thread.

MSDS. Material Safety Data Sheet.

MSHA. Mine Safety and Health Administration.

MSS. Manufacturers Standardization Society of the Valve Fittings Industry, Inc.

MSS Point. Minimum Stable Signal Point; the best superheat setting which will provide constant, or little change in temperature, at the thermostatic expansion valve temperature sensing element while the system is running.

Muck. 1. Mud rich in humus. 2. Finely blasted rock, particularly from underground.

Mud. 1. A slang term for mortar. 2. A slang term for gypsumboard joint compound products. 3. Any soil containing enough water to make it soft. 4. In rotary drilling, a mixture of water with fine drill cuttings and added material which is pumped through the drill string to clean the hole and cool the bit.

Mud Board. See Mortar Board.

Mud Cracks. The cracks that develop in an emulsion coating that has been applied too thickly.

Mud Jacking. Raising a sunken concrete slab by pumping a slurry under the slab through a hole in the slab.

Mud Pan. A hand-held container for holding a small quantity of gypsumboard joint compound products; usually the size of a bread pan.

Muffler, Compressor. Sound absorber chamber in refrigeration system; used to reduce sound of gas pulsations.

Mulch. A mixture, as of leaves and compost, that covers or is mixed with the earth, often to help enrich the soil; bark, crushed stone or other material used to cover planting beds, retain moisture, reduce weeds, and improve appearance.

Mullion. A vertical or horizontal bar between adjacent window or door units; the member between the openings of a window frame to accommodate two or more windows.

Mullion Heater. Electrical heating element mounted in a door or window mullion; used to keep mullion from sweating or frosting.

Mullite. A rare mineral of theoretical composition, a relatively stable phase in ceramics produced by the high temperature reaction of alumina and silica or by the thermal decomposition of alumina-silica minerals such as kyanite, sillimanite, andalusite, and various clay minerals.

Mullite, Porcelain. See Porcelain Mullite.

Multi-Color Spraying. Spraying a surface with two or more different colors at one time from one gun. The multiple colors exist separately within the material and when sprayed create an interlacing color network with each color retaining its individuality.

Multi-Family Dwelling. A dwelling that will accommodate two or more families.

Multi-Layer. Two or more layers of gypsum board used in an assembly.

Multi-Level. A carpeting texture or design created by different heights of tufts, either cut or uncut loop.

Multi-Outlet Assembly. A type of surface or flush electrical raceway designed to hold conductors and attachment plug receptacles, assembled in the field or at the factory.

Multiple Chemical Sensitivity. A condition in which a person is considered to be sensitive to a number of chemicals at very low concentrations.

Multiple Listing. A cooperative sales tool among brokers whereby they share their listings and, in the event of a sale, the listing broker and the selling broker share the commission.

Multiple Stage Compressor. See Compressor, Multiple Stage.

Multiple System. Refrigerating mechanism in which several evaporators are connected to one condensing unit.

Multiport Valve. 1. Filter control valve changing direction of water flow. 2. A valve for various pool filter operations, which combines in one unit the function of two or more single direct flow valves.

Multi-Unit Wall A wall composed of two or more wythes of masonry.

Multi-Zone Air Handling System. A system providing conditioned air similar to a single-zone system; the temperature and flow of the air supplying each zone is controlled separately; a constant supply of air is supplied to the various zones.

Muntin. 1. A small vertical or horizontal bar between small lights of glass in a sash. 2. The vertical member between two panels of the same piece of panel work.

Muntz Metal. An alloy of copper and zinc that contains 60 percent of copper, can be rolled hot, and is used for sheathing and bolts.

Mural. 1. A painting executed directly on a wall or ceiling. 2. Of a wall. 3. Tile installed in a precise area of a wall or floor to provide a decorative design or picture. 4. Glass or marble mosaic tile (tesserae) made to form a picture or design. 5. Ceramic tile, painted and fired to form a picture or design.

Muriatic Acid. Dilute hydrochloric acid (30% HCL), commercial grade, commonly used for cleaning mortar and grout stains from masonry surfaces.

Murphy Bed. A bed that may be folded or swung into a closet.

Murphy's Law. An observation that anything that can go wrong will go wrong.

Museum. A building used for storing and exhibiting objects of historical, scientific, or cultural interest.

Mushroom Capital. A flaring conical head on a concrete column.

Mutual Assent. An objective manifestation by parties that they intend to be bound by a contract.

Mutuality. The concept that a contract, to be enforceable at all, must be enforceable by both parties.

MW. Moderate Weather; see Grade MW Brick.

Mylar. Plastic sheeting used as a drafting medium by drafters.

N

Nadir. The lowest point.

Nail. A stiff metal wire fastening device with a point on one end and head designed for impact on the other end.

Nailable Concrete. Concrete, usually made with suitable lightweight aggregate, with or without the addition of sawdust, into which nails can be driven.

Nailable Studs. Light gauge metal studs with slots that will accept nails for attachment of wall finishes.

Nail Apron. A leather or fabric pouch attached to a carpenter's belt to carry nails or other fasteners.

Nail Bar. A tool which pries open or pulls out driven nails.

Nail Couple. A pair of nails that will resist a rotating force.

Nailer. 1. A wooden strip attached to a concrete, masonry, or steel deck to allow roofing materials to be mechanically fastened. 2. A wooden strip cast into a concrete member for later fastening of finishes by nailing.

Nail Fracture. See Hammer Fracture.

Nail Gun. A mechanical device for driving nails by compressed air.

Nailing Channel. Fabricated from not lighter than 25 gauge steel so as to form slots to permit attachment of lath by means of ratchet-type annular nails, or other satisfactory attachments.

Nailing Pattern. The arrangement, spacings, and dimensions of nails used to attach two or more structural members together.

Nail Popping. Protrusion of the nail from gypsumboard, usually attributed to the shrinkage of or use of improperly cured wood framing.

Nail Pull Resistance. The ability of gypsum board to resist nail head penetration as defined in ASTM C 473.

Nail Set. A small hand tool, like a punch, used for setting finish nails slightly below the surface of the wood.

Nail Spotter. In gypsumboard finishing, a small box-type applicator for covering dimpled nail heads with joint compound; see Ames Taping Tools.

Nameplate Plaque. A flat thin piece of metal used for the inscription of a name.

Nano-. A prefix meaning one billionth of a part, 10^{-9}.

Nanogram. One billionth of a gram.

Nanometer. One billionth of a meter.

Nanosecond. One billionth of a second.

Nanotechnology. The technology of measuring and manufacturing objects of microscopically small size.

NAP. The pile on the surface of a carpet or rug.

Napkin Dispenser. A hollow, box-like metal device installed in women's restrooms to dispense sanitary napkins and tampons.

Naptha. Any of several volatile, flammable liquids obtained by distilling certain materials containing carbon; used as a solvent or thinner in varnish and as a fuel; petroleum naptha is also known as benzine.

Narrow Carpet. Fabric woven 27 in. and 36 in. in width, as distinguished from broadloom.

Narthex. A vestibule leading to the nave of a church.

Natural Asphalt. Asphalt occurring in nature which has been derived from petroleum by natural processes of evaporation of volatile fractions leaving the asphalt fractions; the native asphalt of most importance are found in the Trinidad and Bermudez Lake deposits; asphalt from these sources is called Lake Asphalt.

Natural Clay Tile. A tile made by either the dust-pressed method or the plastic method, from clays that produce a dense body having a distinctive, slightly textured appearance.

Natural Cleft Slate. Slate which has been split into thinner pieces along its natural cleft or seam; it is rougher in appearance than machined slate.

Natural Cleft. A natural V-shaped channel, space, opening, or fissure in a material.

Natural Convection. Fluid motion of air caused by the temperature difference between the solid surface and the fluid with which it is in contact.

Natural Draft. A natural stream of air up a chimney, caused by the stack effect, which draws combustion air through the burner or fuel bed.

Natural Finish. A transparent finish, usually a drying oil, sealer or varnish, applied to wood for protection against soiling or weathering; such a finish should not seriously change the original color of the wood or obscure its grain pattern.

Natural Frequency. The constant frequency of a vibrating system in the state of natural oscillation.

Natural Gas Meter. A mechanical measuring and recording device of the volume of natural gas passing a given point.

Natural Gray Yarn. Unbleached and undyed yarn spun from a blend of black, brown or gray wools.

Natural Resins. Essentially the result of exudation of trees; divided into two large classes, dammars and copals; resins are usually named after the locality in which they are found or the port of shipment.

Natural Stone. Any type of stone which is quarried and not manufactured.

Naval Stores. A term applied to the oils, resins, tars, and pitches derived from oleoresin contained in, exuded by, or extracted from trees, chiefly species of pines; historically, these were important items in the stores of wood sailing vessels.

Nave. The main part of the interior of a church.

NBP. A membrane typically reinforced with polyester; resists tears, punctures and weathering; remains flexible at low temperatures and has a low vapor permeability.

Neat. 1. With no dilution and nothing added. 2. Plaster material requiring the addition of aggregate.

Neat Cement Grout. A fluid mixture of hydraulic cement and water, with or without admixture; also the hardened equivalent of such mixture.

Neat Cement Mortar. Portland cement and water only, no aggregate.

Neat Cement Paste. A mixture of hydraulic cement and water, both before and after setting and hardening.

Neat Cement. Hydraulic cement in the unhydrated state.

Needle Beam. A steel or wood beam threaded through a hole in a bearing wall and used to support the wall and its superimposed loads during underpinning of its foundation.

Needlepoint Valve. Type of valve having a needle point plug and a small seat orifice for low flow metering.

Needle Valve. Similar to a globe valve, it has a needle which seats into a small opening to control the flow of liquid.

Negative Float. The condition in a CPM schedule when there is not enough time available to perform some operation.

Negative Moment. A result of bending moment in a beam in which the upper part is in tension and the lower part is in compression; compare with Positive Moment.

Negative Pressure. 1. Condition that exists when less air is supplied to a space than is exhausted from the space, so the air pressure within that space is less than that in surrounding areas. 2. A pressure within a pipe that is less than atmospheric pressure; also called a Minus Pressure.

Neglect. 1. To pay insufficient attention to. 2. To disregard a duty, leaving it undone.

Negligence. The failure to exercise due care

Negotiable Instrument. A written promise or request for the payment of a certain sum of money to order or bearer; includes checks, bills of exchange, letters of credit, promissory notes, trade acceptances, and certain bonds.

Negotiated Contract. A construction contract where the price has been arrived at by negotiation between owner and contractor rather than by competitive bidding.

Neighborhood. In community planning, a residential area in which residents are within walking distance of each other.

Neighborhood Shopping Center. A small shopping center; a convenience center.

Nelson Stud. A bolt-like rod, sometimes threaded to receive a nut, welded to a steel member for attaching another member.

Neon. An inert gaseous element occurring in traces in the atmosphere and giving an orange glow when electricity is passed through it in a sealed low-pressure tube; used in advertising and commercial signs.

Neoprene. A synthetic rubber (polychloroprene) that is used in fluid or sheet applied elastomeric roofing membranes or flashing; sometimes called chloroprene rubber; sheets are available with or without reinforcing fabric; resists weather, heat, oils, solvents and abrasion; can be used in an inverted roof assembly, a loosely laid, ballasted, or fully adhered.

Neoprene Joint. A fitting of synthetic rubber, neoprene, where two or more members meet.

Neoprene Membrane. An impervious, oil-resistant synthetic rubber, manufactured to be installed in layer form; used for roofing and waterproofing.

Neoprene Roof. A roof system covered with a sheet of synthetic rubber.

Nepheline Syenite. A mineral aggregate consisting chiefly of albite, microcline, and nephelite, each in significant amount.

Net Cross Sectional Area. Of concrete masonry units, the gross cross-sectional area of a section minus the area of cores or cellular spaces; the cross- sectional area of grooves in scored units is not deducted from the gross cross-sectional area to obtain the net cross-sectional area.

Net Effective Temperature (ET). This is not an actual temperature in the sense that it can be measured by a thermometer; it is an experimentally determined index of the various combinations of dry-bulb temperature, humidity, radiant conditions, and air movement that induce the same thermal sensations.

Net Income. The remainder of revenues after direct and indirect costs are deducted; profit.

Net Lease. A lease in which the tenant pays certain specified costs, such as taxes, insurance, and maintenance, of operating the property; a net, net lease and a net, net, net lease are variations where the tenant pays more of the costs.

Net Listing. A real estate listing where the net proceeds of a sale or lease to the owner are specified and all above that figure is the broker's commission; this is an illegal arrangement in some states.

Net Profit. The excess of income after costs and expenses of a business are deducted.

Net Rentable Area. The amount of a building that is available for rental to tenants, expressed in square feet.

Net, Safety, See Safety Net.

Net, Tennis. See Tennis Net.

Net Working Capital. Current assets less current liabilities.

Net Worth. The difference between total assets and total liabilities.

Neutral. 1. Without color; achromatic. 2. Being neither acid nor alkaline. 3. Not electrically live. 4. Neutral Arbitrator.

Neutral Arbitrator. An arbitrator not controlled by or biased in favor of any party; also called a Neutral.

Neutral Axis. The line on a member cross-section on which the bending moment is zero.

Neutral Conductor. The neutral conductor is a grounded conductor; the insulation on a neutral conductor is colored white or a natural gray.

Neutral Flame. A gas flame wherein the portion of oxygen and the welding gas used is neither oxidizing nor reducing.

Neutralizer. Substance used to counteract acids in refrigeration system.

Neutron. That part of an atom core which has no electrical potential; electrically neutral.

New England Method. See Pick and Dip.

Newel. 1. The principal post of the foot of a staircase. 2. The central support of a winding flight of stairs.

Newton. An SI unit of force that is of such size that under its influence a body whose mass is one kilogram would experience an acceleration of one meter per second per second.

NGR Stain. See Non-Grain-Raising Stain.

NIC. Not in Contract.

Niche. Either a curved or square recess in a wall, dependent on the architecture and the use for which it is intended; used for housing statues, vases, telephones, or door chimes.

Nickel. A malleable ductile silver-white metallic element, used in special steels, magnetic alloys, and as a catalyst.

Nickel Silver. A silver-white alloy of copper, zinc, and nickel.

Nickel Steel. A steel alloy containing nickel as its principal alloying element.

Night Depository. An opening in an exterior wall of a structure or building that allows goods or materials to be delivered to a business after it has closed.

NIMBY. Not In My Back Yard, said in opposition to proposals to establish desirable developments, such as schools or hospitals, too close to one's own property.

Nineteen-Inch Selvage. A prepared roofing sheet with a 17 inch granule-surfaced exposure and a nongranule-surfaced 19 inch selvage edge; this material is also called a Split Sheet, SIS or as Wide-Selvage Asphalt Roll Roofing Surfaced with Mineral Granules.

Nipple. Short length of pipe with male threads on both ends.

Nipple, Bushed. See Bushed Nipple.

Nipple, Offset. See Offset Nipple.

NIST. National Institute of Standards & Technology.

Nitrocellulose. Used extensively in making lacquer, is prepared from cellulose (cotton linters) by treatment with chemicals; see Cellulose Nitrate.

Nitrogen Dioxide. Mildly poisonous gas (NO_2) often found in smog and automobile exhaust fumes.

Noble Gas. Any gaseous element of a group that almost never combine with other elements, consisting of helium, neon, argon, krypton, xenon, and radon; also called the Inert Gases or the Rare Rases.

No-Fines Concrete. A concrete mixture in which only the coarse gradation (3/8 to 3/4 inch normally) of aggregate is used.

No-Frost Freezer. Low-temperature refrigerator cabinet in which no frost or ice collects on freezer surfaces or materials stored in cabinet.

Nogging. The filling of brick between the roof rafters from the wall plate to roof boards for making the building wind tight.

No-Hub Pipe. Pipe usually manufactured of cast iron, which is fabricated without hubs and coupled together by a fastener of stainless steel and rubber.

Noil. A by-product in worsted yarn manufacture, consisting of short wool fibers, less than a determined length, which are combed out.

Noise. Any undesired sound.

Noise Reduction Coefficient (NRC). A relative numerical expression of the ability of a material to absorb sound.

Noise Reduction. 1. The reduction in sound pressure level caused by making some alteration to a sound source. 2. The difference in sound pressure level measured between two adjacent rooms caused by the transmission loss of the intervening wall.

Nomenclature. Systematic naming; the terminology of a science.

Nominal Dimension. 1. The named size; compare with Actual Size. 2. An approximate rough cut dimension assigned to a piece of material as a convenience in referring to the piece, such as 2 by 4 or 500 pound. 3. The stated size of a piece of lumber; its actual size is smaller. 4. A dimension greater than a specified masonry unit dimension by the thickness of a mortar joint. 5. This is the approximate facial size or thickness of tile, expressed in inches or fractions of an inch, for general reference, such as a "4 by 4" tile. 6. Tubing measurement which has an inside diameter the same as iron pipe of the same stated size.

Nominal Size. The named size, for example the nominal size of a wood stud is 2- by 4- inches while the actual size is less. The nominal size of a concrete block is 4- by 8- by 16-inch while the actual size is less.

Nomogram. A graphic representation consisting of three or more lines calibrated to scale and arranged so that a straightedge connecting known values on two lines will intersect unknown values on one or more other lines.

Non-Agitating Unit. A truck mounted unit for transporting ready-mixed concrete short distances, not equipped to provide agitation (slow mixing) during delivery.

Non-Automatic. Non-automatic means that the implied action requires personal intervention for its control; see Automatic; as applied to an electric controller, non- automatic control does not necessarily imply a manual controller, but only that personal intervention is necessary.

Nonaxial. In a direction not parallel to the long axis of a structural member.

Nonbearing. Not carrying a load.

Nonbearing Partition. A partition which extends from floor to ceiling but which supports no load other than its own weight.

Non-Bearing Wall. A wall that carries no superimposed load; it could be removed without any structural effect on the remaining building.

Non-Code Installation. Functional refrigerating system installed where there are no local state or national refrigeration codes in force.

Noncombustible. Any material which will neither ignite nor actively support combustion in air at a temperature of 1200° F. when exposed to fire.

Noncondensable Gas. Gas which does not change into a liquid at operating temperatures and pressures.

Nonconforming Use. Use of land that is not in conformance with the zoning code; applies mainly to pre-existing uses when new zoning is enacted; the non- conforming use may remain as long as it is continuous; upon interruption, the non-conforming use may not continue.

Non-Destructive Testing. Testing of building materials in a manner that does not destroy the material.

Non-Drying Oils. Oils which are unable to take up oxygen from the air and change from a liquid to a solid state; mineral oils are non-drying oils.

Non-Evaporable Water. The water in concrete which is irremovable by over drying; chemically combined during cement hydration.

Non-Ferrous Metal. A metal that does not contain iron.

Nonflammable. Incombustible, will not burn; opposite of flammable or inflammable.

Nonfrosting Evaporator. Evaporator which never collects frost or ice on its surface.

Non-Grain-Raising Stain. Wood stain which does not raise the grain of the wood; NGR Stain; made by dissolving dyes such as used in making water stains in special solvent instead of water.

Non-Load Bearing Partition. A structurally non-essential interior wall assembly for compartmentalizing floor space.

Nonmetallic Non-Shrink Grout. Nonmetallic cementitious or epoxy based mix used to fill gap created between bearing components or base plates and foundation or other supporting element.

Non-Metallic Tubing. A round sheath product, of round cross-section, fabricated from a moisture-resistant, flame-retardant material.

Non-Pressure Drainage. Means gravity flow drainage in which the sloping pipes are not filled to capacity.

Nonrecourse Mortgage. One in which the borrower has no personal liability, the lender's sole remedy being to foreclose on the property in the event of a default.

Non-Rising Pin. See Fast Pin Hinge.

Non-Shrink Grout, Nonmetallic. Nonmetallic Non-Shrink Grout.

Non-Shrink Grout. Cementitious or epoxy based mix used to fill gap created between bearing components or base plates and foundation or other supporting element.

Non-Slip Terrazzo. A mixture of marble chips and portland cement on a portland cement and sand body, which has been roughened with iron fillings, carborundum powder or indented while it is wet.

Non-Slip Tile. Tile having greater non-slip characteristics due to an abrasive admixture, abrasive particles in the surface, grooves or patterns in the surface or because of natural non-skid surface characteristics.

Nontoxic. Not poisonous.

Nonvitreous. Non-vitrified; that degree of vitrification evidenced by relatively high water absorption, usually more than 10 percent water absorption, except for floor and wall tile which are considered non-vitreous when water absorption exceeds 7 percent.

Non-Volatile. Portion of a product which does not evaporate at ordinary temperature.

Norm. 1. The standard or principle set by the average of a large group in the same trade or profession. 2. An authoritative standard or model.

Normal Charge. Thermal element charge which is part liquid and part gas under all operating conditions.

Normal Consistency. Of gypsum plaster or gypsum concrete, the number of milliliters of water per 100 grams of gypsum plaster or gypsum concrete required to produce a mortar or a slurry of specified fluidity.

Normal Fault. A fault under tension where the overlying block moves down the dip or slope of the fault plane.

Normal Loading. A design load that stresses a member or fastening to the full allowable stress tabulated; this loading may be applied for approximately 10 years, either continuously or cumulatively, and 90 percent of this load may be applied for the remainder of the life or the member or fastening.

Normal Weight Concrete. Concrete having a hardened density of approximately 150 pcf which is made from aggregate of approximately the same density.

Norman Brick. A brick with 2-3/4 by 4 by 12 inch dimensions.

North Pole. 1. The northernmost point of the earth's axis of rotation 2. The pole of a magnet that points to the north.

Nosing. 1. The projecting forward edge of a stair tread; the part of a stair tread which projects over the riser. 2. Any similar projection. 3. The rounded edge of a board.

Nosing, Rubber. See Rubber Nosing.

Not in Contract (NIC). Something shown on contract drawings, usually for reference, but not to be included in the construction contract.

Notary Public. One authorized by law to acknowledge and certify documents and signatures; a notary.

Notch. A crosswise rabbet at the end of a board.

Notched Trowel. 1. A tile setter's tool consisting of a trowel with serrated edges; used to spread bonding materials; depth of application is regulated by the size and spacing of the teeth. 2. A similar tool used by applicators of resilient flooring to spread adhesive.

Note. A written promise to pay a certain amount of money at a certain time.

Note Payable. Amount owed because of a promissory note given to a creditor.

Note Receivable. A promissory note given by a customer to pay a definite sum to the business at a specific future time.

Notice. Delivery of information to a party

Notice of Cessation. A written notice filed by the owner or owner's agent signifying that no labor has been performed on the project for a specific period of time (such as 60 days) and establishing the time remaining to exercise lien rights by all parties concerned.

Notice of Completion. A written notice filed by the owner or owner's agent signifying completion of a project and establishing the time remaining to exercise lien rights by all parties concerned.

Notice of Nonresponsibility. A notice which, if properly recorded and posted on the premises, relieves the owner from the effects of mechanics liens for work and materials not ordered by the owner, usually ordered by lessees.

Notice to Pay Rent or Quit. A legal notice to a defaulting tenant, a prerequisite to eviction.

Notice to Proceed (NTP). A written notice to a contractor to proceed with the work of the contract.

Novation. The substitution of a new legal obligation for an old one.

Nozzle. An attachment at the end of a plastering machine delivery hose, which regulates the fan or spray pattern.

NRC. Noise Reduction Coefficient.

NRP. Non-Rising Pin.

NTP. Notice to Proceed.

NTS. Not to Scale.

Nuclear Meter. An instrument for detecting and measuring moisture.

Nuisance. A condition of or on real property that damages neighboring persons or property.

Nurse Station Indicator. A signaling device at a nurses station activated by patients needing assistance from nurses.

Nurse Station. An area in a hospital where nurses congregate and medical materials and records are stored.

Nursing Care Facility. Any institution which provides nursing care services to persons needing constant care and attention on a 24-hour basis; a nursing home, nursery, long term facility, convalescent home.

Nut. A small square or hexagonal flat piece of metal or other material with a threaded hole through it for screwing on the end of a bolt to secure it.

Nut, Hex. See Hex Nut.

NW. No Weather; see Grade NW Brick.

Nylon Carpet. A carpet made from synthetic nylon fibers.

Nylon Plastics. Plastics based on resins composed principally on long-chain synthetic polymeric amide which has recurring amide groups as an integral part of the main polymer chain.

Nylon. Any of various synthetic polyamide fibers with tough, lightweight, elastic properties, used in industry and for textiles.

O

Oak. A strong, hard, heavy wood.

Oak Floor. A common type of flooring, usually installed in small tongue and groove strips made of oak.

Oakum. Loosely twisted jute or hemp fiber impregnated with tar and used in caulking seams, as in wood shipbuilding or in caulking cast iron bell and spigot piping.

Oak Veneer. A thin layer of oak glued to an interior wood or bonded together to form a protective or ornamental facing; types available are white oak and red oak.

Oatmeal Paper. Wallpaper made by sprinkling sawdust over adhesive surface.

OB. Octave Band.

Obelisk. A tall, slender, tapering, usually 4-sided, stone monument surmounted by a pyramid, of Egyptian origin.

Oblique. 1. An angle; neither parallel or perpendicular. 2. Not symmetrical with its base.

Oblique Slip Fault. A combination of normal and slip or thrust and slip faults whose movement is diagonal along the dip of the fault plane.

Obscure Glass. Ground glass or frosted glass that transmits light but does not allow a view of objects on the other side.

Obsolescence. The loss of value due to being no longer useful.

Obtuse Angle. An angle of over 90 degrees but less than 180 degrees.

Obtuse Triangle. A triangle having one angle over 90 degrees.

OC. 1. On center; the designation of spacing for framing members in a building from the center of one member to the center of the next; also shown as O/C. 2..See Oxygen Cutting.

Occupancy. A building code term referring to the use of a building, such as school, office, restaurant, warehouse, or residence.

Occupancy Importance Factor (I). A building code term used to identify occupancies of importance.

Ochre. An earth pigment, usually red or yellow, made from impure iron ore, used in paint.

Octave. Frequency difference between harmonic vibrations; it is the doubling of the frequency of sound.

Octyl Alcohol-Ethyl Hexanol. Additive in absorption machines to reduce surface tension in the absorber.

Oculus. 1. A circular opening or eye in a dome. 2. Any round window.

OD. Outside Diameter.

Odor. That property of air contaminants that affect the sense of smell.

Off Cycle. That part of a refrigeration cycle when the system is not operating.

Offer. A promise that is enforceable if accepted.

Offeror. A person who makes an offer.

Office. A room or building where business activities are conducted.

Office Building. A building containing business offices.

Office Cubicle. A small compartment unit, surrounded by fixed or movable walls on two or three sides, where an individual works.

Office Park. A planned, usually suburban, community of office buildings with common management, parking, landscaping, and other services.

Office Partition. Moveableinterior wall, used to partition off an office area, assembled together to create cubicles for employees.

Office Safe. A locked receptacle that holds money or valuables in an office; usually rated by the manufacturer to designate fire-worthiness and strength.

Office Trailer. A highway vehicle parked on a job site, designed to serve wherever needed, as an office and a place to carry out business.

Offset. 1. A combination of elbows or bends which brings one section of pipe out of line and back in line parallel with the original sector. 2. A course of masonry units that sets in from the course that is directly below it; also called Set Back, Set In, or Set Off; the opposite of Corbel.

Offset Cabinet Construction. A method of building cabinets where the back corners of the doors and drawer fronts are rabbeted so they overlap the face frame, thereby concealing the joint.

Offset Connector. A fitting that is a combination of elbows or bends which brings one section of pipe out of line with, but into a line parallel with, another section.

Offset Hinge. A hinge with a 90 degree offset in one or both legs, designed to shift the center of rotation of the door.

Offset Nipple. A fitting, threaded at both ends, that is a combination of elbows or bends which brings one section of pipe out of line with, but into a line parallel with, another section.

Off-Site. Anything not on the subject building site.

OG. Ogee.

Ogee. A curved section of a moulding, partly convex and partly concave.

Ogee Molding. A molding with a modified S-shaped profile.

Ogee Roof. An architectural roof design which is Eastern influenced.

Ogive. A pointed arch.

Ohm (R). Unit of measurement of electrical resistance; one ohm exists when one volt causes a flow of one ampere.

Ohmmeter. Instrument for measuring resistance in ohms.

Ohm's Law. A law in electricity which states the relationship between voltage, amperes, and resistance; equation. Amperes(I) x Ohms(R) = Volts(E).

Oil Absorption. A measure of the ability of pigments to absorb oil.

Oil Base Caulking. Caulking made with resins and other ingredients made of various oils, dispensed through a tube.

Oil Binding. Condition in which an oil layer on top of refrigerant liquid may prevent it from evaporating at its normal pressure-temperature.

Oil Burner Nozzle. A spraying device which atomizes fuel oil into a fine spray and increases the combustion characteristics of the fuel.

Oil-Canning. Describing the distortion of thin-gauge metal panels that are fastened in a manner restricting normal thermal movement.

Oilcloth Varnish. Special highly flexible and elastic varnish.

Oil Colors. Colors ground to form of paste in linseed oil.

Oil Fired Boiler. A boiler that is heated by a unit that sends oil under high pressure to a nozzle, where it is sprayed as a mist and ignited by an electric spark.

Oil Furnace. A furnace that burns heating oil.

Oil Length. Oil length in varnish is measured by the oil in gallons per hundred pounds of resin; a long-oil varnish is tougher than a short-oil varnish; rubbing varnish is a typical short-oil varnish and spar varnish is a typical long-oil varnish.

Oil of Vitriol. See Sulphuric Acid.

Oil Paint. Paint in which a drying oil is the vehicle.

Oil, Refrigeration. See Refrigeration Oil.

Oil Separator. Device used to remove oil from gaseous refrigerant.

Oil Soluble. Capable of being dissolved in oil.

Oil Stain. A wood stain consisting of oil-soluble dyes and solvents such as turpentine, naphtha, or benzol,; penetrates into pores of wood; has tendency to bleed.

Oilstone. A whetstone used with oil.

Oil Switch. A switch or protector relay installed in a flue pipe, that will shut off a burner in case the stack does not come up to a predetermined temperature within 45 seconds after the motor starts.

Oil Tank. The storage canister where heating oil is stored for future use by an oil boiler, or an oil burning furnace system.

Oil Varnish. A varnish consisting of a hard resin combined with a drying oil and a drier thinned with a volatile solvent; after application, the solvent dries first by evaporation; then the oil dries by oxidation.

Oiticica Oil. Drying oil obtained from nut of oiticica tree.

Old English Bond. See English Bond.

Old Growth. Timber in or from a mature, naturally established forest. When the trees have grown during most if not all of their individual lives in active competition with their companions for sunlight and moisture, this timber is usually straight and relatively free of knots.

Olefin. Long-chained synthetic polymer composed of at least 85% by weight of ethylene, propylene or other olefin units; currently, only polypropylene has been produced in fiber form for carpet manufacture.

Olefin Plastics. Plastics based on resins made by the polymerization of olefins or copolymerization of olefins with other unsaturated compounds, the olefins being in greatest amount by weight; polyethylene, polypropylene and polybutylene are the most common olefin plastics encountered in pipe.

Oleoresin. A solution of resin in an essential oil that occurs in or exudes from many plants, especially softwoods; the oleoresin from pine is a solution of pine resin (rosin) in turpentine.

Oleoresinous Varnish. Varnish composed of resin or gum dissolved in a drying oil which hardens as it combines with oxygen from the air.

Olive Knuckle Hinge. A paumelle hinge with knuckles forming an oval shape.

On Center (O/C). Center to center.

One Piece Toilet. A toilet where the water tank and bowl are cast in one piece of vitreous china for the purposes of quieter operation and better appearance.

One-Family Dwelling. A dwelling containing only one dwelling unit.

One-Way Action. The structural action of a slab that spans between two parallel beams or bearing walls.

One-Way Concrete Joist System. A reinforced concrete framing system in which closely spaced concrete joists span between parallel beams or bearing walls.

One-Way Glass. See Transparent Mirror.

One-Way Solid Slab. A reinforced concrete floor or roof slab that spans between parallel beams or bearing walls, with structural reinforcing in the spanning direction only.

One-Way. Constructed with reinforcing steel running in one direction only.

On-Site. Objects and activities that are on the building site.

On the Stump. See Standing Timber.

Opacity. Hiding power, as of paint.

Opaque. Impervious to light; not transparent.

Opaque Glaze. A non-transparent glaze with or without color.

Open Circuit. Electric circuit with a physical interruption caused by a switch, a break in a conductor, or an open breaker or fuse.

Open Compressor. An external drive compressor; not hermetic.

Open Decking. A deck in which the joists on the underside are exposed.

Open Display Case. Commercial refrigerator designed to maintain its contents at refrigerating temperatures even though the contents are in an open case.

Open Front Seat. A toilet seat with an open gap at the front, usually used in public toilet rooms for improved sanitation.

Open-Graded Aggregate. One containing little or no mineral filler and in which the void spaces in the compacted aggregate are relatively large.

Open-Graded Asphalt Friction Course. A pavement surface course that consists of a high-void, asphalt plant mix that permits rapid drainage of rainwater through the course and out the shoulder. The mixture is characterized by a large percentage of one-sized coarse aggregate. This course prevents tire hydroplaning and provides a skid-resistant pavement surface.

Open-Grained Wood. Woods with large pores, such as oak, ash, chestnut, and walnut; also known as Coarse-Grained or Coarse Textured.

Open Listing. A listing agreement with one or more real estate brokers to sell or lease one's property on a non-exclusive basis.

Open Run-Around. A method of heat recovery from exhaust air in which a fluid is inserted alternately in the supply and exhaust air stream; this system transfers both sensible and latent heat.

Open Time. The period of time during which the bond coat retains its ability to adhere to the tile and bond the tile to the substrate.

Open Type Compressor. See Compressor, Open Type.

Open Type Decking. A deck in which the joists on the underside are exposed.

Open-Web Steel Joist. A prefabricated, welded steel truss used at closely spaced intervals to support floor or roof decking.

Operable Door. A door, of a pair, that is normally operable.

Operable Part. A part of a piece of equipment or appliance used to insert or withdraw objects, or to activate, deactivate or adjust the equipment or appliance, such as a coin slot, pushbutton, or handle.

Operating Instructions. The written explanations and descriptions furnished by manufacturers and fabricators of building equipment and components for the proper operation of their products.

Operating Leverage. The ratio of the percentage change in profit to the percentage change in sales.

Operating Pressure. Actual pressure at which a refrigerating system works under normal conditions; this pressure may be positive or negative (vacuum).

Operating Profit. Revenue of the business less the total of all costs of the operation.

Operating Room. A room in a hospital for surgical operations.

Operation Expenses. Total of all expenses including cost of sales, general and administrative expenses as well as indirect operating expenses.

Opposed Blades. Two sets of blades in a damper, linked so that the adjacent blades can open and turn in opposite directions.

Option. 1.Something that may be chosen. 2. A right given for a consideration to keep an offer to purchase or lease open for a specific time.

Optionee. 1. One who receives an option. 2. The prospective purchaser or lessee in a real estate option.

Optionor. 2. One who gives an option. 2. The owner-seller or lessor in a real estate option

Oral Agreement. A contract in words that are not reduced to writing.

Orange Mineral. Red lead prepared by roasting basic carbonate white lead; used mainly in printing ink for its characteristic color.

Orange Peel. 1. A pitted texture of a fired glaze resembling the surface of rough orange peel. 2. Spray painting defect in which the lacquer coat does not level down to a smooth surface but remains rough, like the peeling of an orange.

Orbital Sander. A hand-held platform sander that moves rapidly in a circulatory movement.

Ordinance. A law enacted by a city or a county.

Ore. Rock or earth containing workable quantities of a mineral or minerals of commercial value.

Or Equal. The phrase used in specifications providing that substitutions must be equal in all relevant respects and usually subject to the approval of the architect.

Organic. 1. Any chemical compound containing carbon. 2. Being or composed of hydrocarbons or their derivatives or matter of plant or animal origin. 3. Pertaining to or derived from living organisms.

Organic Adhesive. A prepared organic material, ready to use with no further addition of liquid or powder, used for bonding tile to back-up material by the thinset method; cures or sets by evaporation.

Organosol. Film former containing resin plasticizer and solvent; colloidal dispersion of a resin in plasticizer containing more than 5 percent volatile content.

Oriel Window. A bay window projecting from a wall and supported on corbels or brackets.

Oriented Strand Board (OSB). A type of particle panel product composed of strand-type flakes which are purposefully aligned in directions which make a panel stronger, stiffer, and with improved dimensional properties in the alignment directions than a panel with random flake orientation; also called Strandboard.

Orifice. 1. An opening through which something may pass. 2. Attachment to the nozzle on the hose of a plastering machine, of various shapes and sizes, which may be changed to help establish the pattern of the plaster as it is projected onto the surface being plastered. 3. The opening through which fuel is ejected into the burner unit of a furnace or boiler.

Original Contractor. In the law of mechanics liens, a contractor who contracts directly with an owner of real property.

O-Ring. A rubber seal used around stems of some valves to prevent water from leaking by it.

Ormolu. A gilded bronze or gold-colored alloy of copper, zinc, and tin used to decorate furniture and make ornaments.

Ornament. A thing used or serving to adorn; a decoration added to embellish.

Ornamental. Serving as an ornament; decorative.

Ornamental Facing. In masonry, a design formed by the laying of stone, brick, tile or other masonry units so as to produce a decorative effect.

Ornamental Metal. A detail that is added to a building constructed of metal, with the purpose of embellishment or decorating the structure.

Ornate. Elaborately adorned; highly decorated.

Orthogonal. Of or involving right angles.

Orthographic Projection. Projection of a single view of an object on a drawing surface that is perpendicular to both the view and the lines of projection; floor plans, elevations, and sections of buildings are orthographic projections.

Orthotropic. Having unique and independent properties in three mutually orthogonal (perpendicular) planes of symmetry; a special case of anisotropy.

OS&Y Valve. A type of valve, with external exposed threads supported by a yoke, indicating the open or closed position of the valve.

OSB. Oriented Strand Board.

Oscillation. A repeated back and forth movement, from one extreme to another, periodically from a maximum to a minimum.

Oscillograph. A device for recording oscillations.

Oscilloscope. A cathode ray tube that shows the changing voltage of an electrical signal as a curved line on a screen.

Osmosis. The passage of a solvent through a semi-permeable membrane into a more concentrated solution.

Ottawa Sand. A sand used as a standard in testing hydraulic cements by means of mortar test specimens; produced by processing silica rock particles obtained by hydraulic mining of the orthoquartzite situated in open pit deposits near Ottawa, Illinois; naturally rounded grains of nearly pure quartz.

Outdoor Exposure. Plastic pipe placed in service or stored so that it is not protected from the elements of normal weather conditions, the sun's rays, rain, air, and wind.

Outfall Sewer. A large sewer leading from the lower end of the collecting system to the place of disposal.

Outlet. 1. An electrical receptacle into which appliances may be plugged. 2. A way out; an exit.

Outlet Box. A box or container which houses an electrical outlet and its connections.

Outline Lighting. An arrangement of incandescent lamps or gaseous tubes to outline and call attention to certain features such as the shape of a building or the decoration of a window.

Outline Specifications. An abbreviated form of specifications normally produced with schematic design or design development drawings.

Out of Phase. The state wherein a structure in motion is not at the same frequency as the ground motion; or where equipment in a building is at a different frequency from the structure.

Out of Square. The variation from true square.

Outrigger. A structural member projecting from a main structure to provide additional stability or to support something, such as an outside scaffold.

Outrigger, Flagpole. See Flagpole Outrigger.

Outrigger Scaffold. A work scaffolding supported on outrigger beams cantilevered out from the building or structure being worked on.

Outside Four Inches. The single tier of stretcher courses on the face of the wall.

Oven, Bake. See Bake Oven.

Oven, Convection. See Convection Oven.

Ovendry Wood. Wood dried to a relatively constant weight in a ventilated oven at 102 to 105° C.

Oven, Wall Mounted. See Wall Mounted Oven.

Overatomized. Dispersed too finely by use of excessive atomizing air pressure.

Overcertification. The process of an architect or engineer certifying more than the proper amount to be paid to a contractor.

Overcoat. Second coat; topcoat.

Overcurrent. An electrical current that is in excess of the capacity of the wiring and devices.

Overcurrent Device. Any device that limits the current in a wire to a predetermined number of amperes.

Overflow. An opening to allow rainwater to overflow after it reaches a predetermined height, commonly 2 inches in the case of roof drains.

Overflow System. Surface water collection or draw-off arrangement.

Overglaze Decoration. A ceramic or metallic decoration applied and fired on the previously glazed surface of ceramic ware.

Overhand Work. A complete wall that is built from a scaffold that is located only on one side of the wall; the outside or farther face brickwork is laid by reaching over the wall.

Overhang. 1. The part of a roof that extends beyond the exterior walls of the building. 2. An overhead projection from the wall. 3. A leaning of the wall face away from the vertical.

Overhang Beam. A beam that is supported by two or more supports and has one or both ends projecting beyond the support.

Overhead. All of the indirect costs which can not be directly related to a unit of production and which must be included in the cost of the sale by some allocation method.

Overhead Concealed Closer. A door closer concealed in the head frame with an arm connecting with the door at the top rail.

Overhead Door. A door, commonly used in garages and warehouses, that opens upward from the ground.

Overimprovement. An improvement of excessive size or cost in comparison with similar properties particularly when it is not the highest and best use of the land.

Overlap. 1. The width of fresh paint covered by next layer. 2. Protrusion of weld metal beyond the bond at the toe of the weld.

Overlay. 1. A drawing on a thin layer of paper superimposed over a base drawing. 2. A thin layer of paper, paper, plastic, film, metal foil, or other material bonded to one or both faces of wood panel products or to lumber to provide a protective or decorative face or a base for painting.

Overlay Cabinet Construction. A method of building cabinets where the doors and drawer fronts are in front of the face frame.

Overload. Load greater than that for which system or mechanism was intended or designed.

Overload Protector. Device, either temperature, pressure or current operated, which will stop operation of unit if dangerous conditions arise; see Overcurrent Device.

Oversized Brick. A brick with dimensions greater than 2-1/2"x 3-1/2"x 7-1/2".

Overspray. See Drift, 2.

Overvibration. Excessive vibration of freshly mixed concrete during placement causing segregation.

Ovolo. A quarter round molding.

Owner. 1. Any person or entity having some title or interest in a parcel of real property. 2. A person or entity who contracts for construction.

Owner-Builder. An owner of property who undertakes to construct improvements without the services of a general contractor.

Owner's Overall Budget. A budget prepared by the owner of all of the project costs including land acquisition, construction, financing, professional fees, and all other incidental costs of the project.

Owner's Right to Stop Work. The owner has the right to stop the work of a construction contract only if the contract so provides and only for the reasons stated therein.

OX, OXO, OXXO. Designations of the arrangement of sliding and fixed panels in sliding glass doors and windows; X indicates a sliding panel; O indicates a fixed panel; see also XO, XOO, XOOX.

Oxalic Acid. A poisonous strong organic acid, usually a solid white granular substance; used to dissolve iron rust stains on pool walls and floors, to clean iron rust from filter septa, and as a wood bleach.

Oxidation. The process of combining with oxygen.

Oxidize. To combine with oxygen; to become rusty.

Oxidized Sewage. Sewage which has been exposed to oxygen which causes organic substances to become stable.

Oxidizing Flame. A gas flame wherein the portion used has an oxidizing effect by an excess of oxygen.

Oxy-Acetylene. A mixture of oxygen and acetylene, used as a fuel in cutting or welding metals.

Oxygen. A colorless tasteless odorless gaseous element, occurring naturally in air, water, and most minerals and organic substances, and essential to plant and animal life.

Oxygen-Acetylene Cutting. An oxygen-cutting process wherein the severing of metals is effected by means of the chemical reaction of oxygen with the base metal at elevated temperatures, the necessary temperature being maintained by means of gas flames obtained from the combustion of acetylene with oxygen.

Oxygen-Acetylene Welding. A gas welding process wherein coalescence is produced by heating with a gas flame or flames obtained from the combustion of acetylene with oxygen with or without the application of pressure and with or without the use of filler metal.

Oxygen-Arc Cutting. An oxygen-cutting process where in the severing of metals is effected by means of the chemical reaction of oxygen with the base metal at elevated temperatures the necessary temperature being maintained by means of an arc between an electrode and the base metal.

Oxygen Cutting (OC). A group of cutting processes wherein the severing or removing of metals is effected by means of the chemical reaction of oxygen with the base metal at elevated temperatures; in the case of oxidation-resistant metals the reaction is facilitated by the use of a chemical flux or metal powder.

Ozalid. Tradename of a system of drawing reproduction based on the ammonia vapor process; the prints consist of purple lines on a white background; other colors of line and background are also available.

Ozone. A form of oxygen, O_3, a faintly blue unstable gas with a pungent odor and powerful oxidizing properties; used in disinfection, deodorization, oxidation, and bleaching.

P

Packaged Chiller. A factory assembled piece of equipment that utilizes a refrigeration cycle to produce chilled water for circulation to a desired location or use.

Package Deal. An offering of design, construction, financing, and sometimes the land, for one all-inclusive price.

Packaged Heater. A factory assembled piece of equipment that supplies heat producing units for circulation to a desired location or use.

Packaged Terminal Air Conditioner (PTAC). A self contained unit, operated in the direct expansion method, located in the space served.

Package Dyeing. Placing spun and wound yarn on large perforated forms and forcing the dye through the perforations.

Package Units. Complete refrigerating system including compressor, condenser and evaporator located in the refrigerated space.

Packing. Sealing device consisting of soft material or one or more mating soft elements; reshaped by manually adjustable compression to obtain or maintain a leak-proof seal.

Packing House Tile. Similar to quarry tile but usually of greater thickness.

Pad, Bearing. See Bearing Pad.

Pad, Equipment. See Equipment Pad.

Pad Eyes. Metal rings mounted vertically on a plate for tying small vessels.

Padding, Carpet. See Carpet Padding.

Pad, Transformer. See Transformer Pad.

Pagoda. A multi tiered tower of India and the Far East erected as a temple or memorial.

Paint. A mixture of pigment, binder, and solvent as a suitable liquid to form a thin closely adherent coating when spread on a surface in a thin coat to decorate or adorn by applying lines and color.

Paint, Asphaltic. See Asphaltic Paint.

Paint Brush. A painter's implement composed of bristles set into a handle, used for applying paint and other coatings to a surface.

Paint Coating. Paint in position on a surface.

Paint, Door. See Door Paint.

Painter. A craft worker skilled in the application of paints, stains, and other decorative and protective treatments.

Paint Gauge. An instrument used to measure the thickness of paint coatings.

Paint Heater. Device for lowering viscosity of paint by heating.

Painting. The work done by painters.

Painting Preparation. The act or process preparing a building or structure for painting, including scraping, filling, sanding, and mixing of paint.

Paint Preparation. The act or process of preparing a building or structure for painting, including filling, scraping, sanding, and mixing of paint.

Paint Project. Single paint job.

Paint Remover. A mixture of active solvents used to remove paint and varnish coatings.

Paint Shop. An area in a building being painted where the painter stores, mixes, cleans tools, and prepares painting materials and equipment.

Paint Spray Booth. A room or enclosed space that provides a dust-free environment and ventilation for the application of paints.

Paint System. The complete number and type of coats comprising a paint job, including surface preparation, pretreatments, dry film thickness, and manner of application.

Palladian. In the neo-classical style of Andrea Palladio.

Pallet. A portable platform for storing, moving, and handling goods and materials as in a warehouse.

Pan. Metal or plastic prefabricated form unit used in construction of concrete floor joist systems.

P and L. Profit and Loss.

Pane. A panel of glass.

Panel. 1. A distinct, usually rectangular, section of the surface of a door, cabinet, wall, ceiling or roof. 2. A board containing instruments, controls, dials, and switches. 3. A group of people, usually three, forming an arbitration tribunal. 4. A large, thin board or sheet of lumber, plywood or other material. 5. A thin board with all its edges inserted in a groove of a surrounding frame of thick material. 6. A section of floor, wall, ceiling or roof, usually prefabricated and of large size, handled as a single unit in the operations of assembly and erection. 7. Form section consisting of sheathing and stiffeners that can be erected and stripped as a unit. 8. A recessed section on the broad face of a brick; see Frog.

Panel, Acoustical. See Acoustical Panel.

Panelboard. A single panelboard or group of panel units designed for assembly in the form of a single panel including buses and with or without switches or automatic overcurrent protective devices, or both, for the control of light, heat, or power in a cabinet or enclosure placed in or against a wall or partition and accessible only from the front; also called a Switchboard.

Panel Circuit Directory. A systematic listing of all of the circuits in an electric panel.

Panel Cladding. Metal sheathing panels used to provide durability, weathering and corrosion, or impact resistance.

Panel Edge Clip. H-shaped clip used to prevent differential deflection of plywood roof or floor sheathing.

Panel, Hardboard Siding. See Hardboard Panel Siding.

Panel Hoist. A device to raise sheetrock into place while it is being nailed or screwed.

Paneling. 1. Panels joined in a continuous surface. 2. Decorative wood panels.

Panelized Roof. A modular roof framing system so dimensioned that the sheathing and rafters can be prefabricated as a unit for efficient installation on the purlins or beams.

Panel, Limestone. See Limestone Panel.

Panel Rated Siding. Panel siding made of APA proprietary siding product.

Panel, Sandwich. See Sandwich Panel.

Panel Siding. Large sheets of plywood or hardboard which may serve as both sheathing and siding.

Panic Device. A mechanical device that opens a door automatically if pressure is exerted against the device from the interior of the building; a horizontal bar mounted across the full width of a door or a sort of large push-plate, acting as a latching system and operated by inside pressure on the bar or plate; also called Panic Hardware.

Panic Hardware. See Panic Device.

Pan, Metal Stair. See Metal Stair Pan.

Pantry. A room for storing food and for serving to the table.

Paper and Wire. Asphalt impregnated paper and wire mesh, or metal lath, that are used as a backing for the installation of tile.

Paper Backed Insulation. Insulation that comes with paper facing on one side which serves as a vapor retarder.

Paper-Base Laminate. (See Laminate, Paper-Base).

Paper, Building. See Building Paper.

Paperboard. The distinction between paper and paperboard is not sharp, but broadly speaking, the thicker (over 0.012 inch), heavier, and more rigid grades of paper are called paperboard.

Paper, Curing. See Curing Paper.

Paper Mounted Mosaics. Ceramic mosaic tiles mounted on paper applied to face of tile; available in sheets approximately 12 by 24 inches.

Paper Rollers. Curlings of paper torn from the surface of the gypsum board; usually occurs as a result of prolonged exposure to high humidity and sliding one board across the surface of another.

Papreg. Any of various paper products made by impregnating sheets of specially manufactured high-strength paper with synthetic resin and laminating the sheets to form a dense, moisture-resistant product.

Para Red. Pigment which is coal tar product; brilliant, opaque, non-fading, but has tendency to bleed.

Parabola. An open plane curve formed by the intersection of a cone with a plane parallel to its side.

Parabolic Troffer. A channel-like enclosure for light sources with a reflector shaped to control the light in a narrow beam.

Paraffin Oil. Light gravity mineral oil used as a lubricant in wood finishing.

Parallel. Lines side by side, equidistant, and never meeting.

Parallel Application. The long dimension of the gypsum board applied in the same direction as the framing members; also called Vertical Application or With Framing.

Parallel Bar. A pair of bars on a support adjustable in height and spacing that are parallel to each other and are used for gymnastic exercises.

Parallel Circuit. Arrangement of electrical devices in which the current divides and travels through two or more paths and then returns through a common path.

Parallel Rule. A device mounted on a drawing board to enable a drafter to produce parallel horizontal lines, and with triangles, to produce parallel vertical or sloping lines.

Parapet Wall. That portion of any exterior wall, party wall, or fire wall which extends above the roof line.

Parapet. The region of an exterior wall that projects above the level of the roof.

Parcel. A tract or plot of land under one legal description.

Parge. 1. To apply ornamental plaster to. 2. To apply waterproofing plaster to.

Parget. Parge.

Pargeting. Parging.

Parging. 1. Ornamental plaster. 2. Application of cement plaster as a part of a waterproofing system, as on foundation walls and rough masonry.

Parking Barrier. A structure either temporary or permanent, that is placed to prevent the encroachment of vehicles.

Parking Stall Painting. The act or process of painting border lines for the parking of individual motor vehicles.

Parol Evidence Rule. The meaning of a contract cannot be changed by reference to statements made by the parties to the contract before the contract was formed.

Parquet Floor. A floor covering laid out in a geometric design composed of small pieces of wood.

Partial Joint Penetration. Joint penetration which is less than complete.

Partial Pressures. Condition where two or more gases occupy a space and each one creates part of the total pressure.

Particleboard Underlayment. Flat sheet material of particleboard placed over subflooring to provide a smooth and even surface to receive finish floor covering.

Particleboard. A building panel composed of small particles of wood and resins bonded together under pressure; lat sheet material producing a durable and dimensionally stable product which is often used in dry conditions in place of plywood.

Particle-Size Distribution. The gradation in size of aggregate particles used in concrete, usually expressed in terms of cumulative percentages smaller or larger than each of a series of sieve openings or percentages between certain ranges of sieve openings.

Parting Bead. A narrow vertical strip in a double-hung window frame separating the upper and lower sashes; also called Parting Stop.

Parting Compound. A substance applied to concrete formwork to prevent concrete from adhering.

Parting Stop. Parting Bead.

Partisan Arbitrator. An arbitrator appointed by, controlled by, or biased in favor of a party to a dispute; a Party-Appointed Arbitrator.

Partition. A permanent interior wall which serves to divide a building into rooms.

Partition, Clay Tile. See Clay Tile Partition.

Partition Closure. Method of construction used to close openings formed between flutes of deck and abutting walls or partitions; resilient foam material is often used for this purpose.

Partition Door. Door located in Wire Mesh Partition.

Partition Fabric. The wire mesh fabric component of Wire Mesh Partition.

Partition Framing Member. Metal framework component of Wire Mesh Partition.

Partition System. An assembly of materials designed to perform a special function as a wall.

Partition, Toilet. See Toilet Partition.

Partition, Wire Mesh. See Wire Mesh Partition.

Partner. One of the owners of a partnership.

Partnering. An educational procedure for obtaining the cooperation and understanding of all concerned in a construction contract by their participation at meetings or seminars attended by owner, architect, engineers, contractors, subcontractors, and suppliers to find out what is important to each participant.

Partnership. A business relationship between two or more persons, based upon a written, oral, or implied agreement, to combine their resources and skill in a joint enterprise and to share profits and losses jointly.

Parts Per Million (PPM). 1. Unit of concentration, as in solutions. 2. In swimming pools, the parts of a chemical or mineral per million parts of water by weight.

Party-Appointed Arbitrator. An arbitrator appointed by one of the parties; see Partisan Arbitrator.

Party Wall. 1. A special purpose wall system used to divide compartments for different occupancies; may have requirements for fire and sound. 2. Partitions of brick or stone walls between buildings on two adjoining properties in which each of the respective owners of the adjacent buildings share the rights and benefits of the common wall. 3. A common wall between two tenancies; also called Demising Wall.

Pascal's Law. A law of hydrostatics that states that in a perfect fluid the pressure exerted on it anywhere is transmitted undiminished in all directions.

Pass. 1. In spray painting, motion of the spray gun in one direction only. 2. A single longitudinal progression of a welding operation along a joint or weld deposit; the result of a pass is a weld bead. 3. A working trip or passage of an excavating or grading machine.

Passage Door. A door other than an exit door through which persons may traverse.

Passenger Elevator. An elevator used for people.

Passivation. Act of making inert or unreactive.

Passive Pressure. The horizontal resistance of the soil to forces against the soil.

Passive System. A solar heating or cooling system that uses no outside mechanical power to move the collected solar heat.

Pass-Through. An opening between two rooms or in an exterior wall through which things may be passed, as in serving food or returning soiled dishes.

Pastel. A light and subdued shade of a color.

Paste Wood Filler. A compound supplied in the form of a stiff paste for filling the open grain of hardwoods, such as oak, walnut, and mahogany.

Pastiche. An architectural design that imitates previous designs or styles; a stylistic imitation; a hodgepodge; see Eclectic.

Patch. A piece of material used to mend or cover a hole or as reinforcement.

Patch, Concrete. Material for the repair of small to medium holes or cracks in concrete surfaces.

Pate Dure. Hard paste, aFrench term designating ceramics fired at relatively high temperatures.

Patent Ambiguity. An expression in a document that is obviously susceptible to more than one interpretation.

Patent-Back Carpet. Carpet so constructed that the fabric can be cut in any direction, without raveling of edges; the edges are joined by tape and adhesives instead of being sewed.

Patent Defect. A defect or shortcoming in a structure that is apparent to reasonable inspection.

Pate Tendre. Soft Paste; a French term designating ceramics fired at relatively low temperatures.

Path of Travel. A passage that may consist of walks and sidewalks, curb ramps and pedestrian ramps, lobbies and corridors, elevators, other improved areas, or a necessary combination thereof, that provides free and unobstructed access to and egress from a particular area or location for pedestrians and/ or wheelchair users.

Patina. A green film formed on copper or bronze over time.

Patio. 1. An enclosed unroofed courtyard. 2. A paved area adjacent to a home, used for recreation or outdoor dining.

Patio Block. Lightweight concrete paving slabs installed in lightly used foot traffic areas.

Pattern. A form, model, or design from which copies can be made.

Pattern Length. In spray painting, length of spray pattern.

Pattern Width. In spray painting, width of spray at vertical center.

Paumelle. A style of hinge embodying a single pin of the pivot type, generally of a smooth, streamlined design.

Pavement. 1. A paved surface. 2. The material used to pave a surface such as a street or parking area.

Pavement Cutting. The process of scoring or cutting through pavement surfaces with a power saw with a specific blade for that purpose.

Pavement Marking. The act or process of applying painted lines or necessary instructional signage on pavement surfaces for pedestrians or vehicle drivers.

Pavement Structure. A pavement structure with all its courses of asphalt-aggregate mixtures, or a combination of asphalt courses and untreated aggregate courses placed above the subgrade or improved subgrade.

Paver. 1. One who paves. 2. Any durable stone, brick, or tile unit suitable for construction of a pedestrian or vehicular surface. 3. A half-thickness brick used as finish flooring. 4. Unglazed porcelain or natural clay tile formed by the dust-pressed method and similar to ceramic mosaics in composition, but thicker. 5. Glazed paver tile.

Paver Brick. Brick units that are used in foot traffic areas; usually four inches wide, eight inches long, and 1-5/8 to 2-1/4 inches thick.

Paver, Granite. See Granite Paver.

Paver, Marble. See Marble Paver.

Paver, Masonry. See Masonry Paver.

Paver, Stone. See Stone Paver.

Paving. The surface of an outdoor area; pavement.

Paving, Concrete. See Concrete Paving.

Paving Finishes. The finish coats of concrete, asphalt or coated macadam on streets, sidewalks, and parking areas.

Paving Unit. A precast masonry unit, usually 2-1/4 inches thick used for stepping stones; patios, veneering, and paving; also called a Cap Block.

Payback Period. 1. The period required to recover the investment in an asset. 2. A method utilized by management to evaluate the profitability of alternative investment proposals.

Payment. Satisfaction, or partial satisfaction, of a debt.

Payment Bond. A guarantee by a surety that those persons who supply work and materials to a construction project will be paid for the work and materials

Payment Schedule. A provision in a construction contract that specifies the times for, and amounts of, payments for construction services.

Payroll. The record of wages, salaries, and fringe benefits paid by an employer to its employees.

PC. 1. Portland cement. 2. Prime coat. 3. Pull chain.

PCB. Polychlorinated Biphenyl.

PCC. Portland Cement Concrete.

P.C. Concrete. Portland Cement Concrete; concrete composed of coarse and fine aggregates, portland cement, water, and sometimes admixtures to impart special qualities to the concrete or to aid in placing or curing.

PCF. Pounds per Cubic Foot.

PCI. 1. Pounds per Cubic Inch. 2. Prestressed Concrete Institute.

PDI. Plumbing and Drainage Institute.

PE. Professional Engineer.

Pea Gravel. That portion of concrete aggregate passing the 3/8 sieve and retained on a No. 4 sieve.

Peak. The top point of a roof, truss, spire, finial or any other similar structure.

Pearl Lacquer. Lacquer into which has been suspended guanine crystals, multi-faceted crystals found in skin attached to scales of sardine herring that thrive in cold water.

Peat. See Humus.

Peat Moss. Sphagnum.

Peavey. A lumber worker's tool for handling logs, consisting of a stout wood handle with a steel hooked lever at one end.

Pecan Veneer. A thin layer of hardwood glued over a core of sturdier less valuable solid wood, or plywood, used in flooring and furniture.

Peck. Pockets or areas of disintegrated wood caused by advanced stages of localized decay in the living tree; usually associated with cypress and incense-ceda; there is no further development of peck once the lumber is seasoned.

Peculiar Risk Doctrine. The doctrine that an owner, by employing an independent contractor, cannot escape liability to persons who may be injured during construction operations on the owner's property, since the construction operations involve a special risk of harm.

Pecuniary. Consisting of or measured in money.

Pedestal. 1.An upright compression member whose height does not exceed three times its average least lateral dimension. 2. A short compression member of reinforced concrete that is placed between a column and the footing to distribute the load to the footing. 3. A support for a column or statue.

Pedestal Floor. A flooring system which has short piers or legs used as a base and the flooring laid over those piers to provide a floor system; special flooring designed to prevent electrostatic buildup and sparking in a computer room; usually elevated over the existing floor to facilitate the installation of wires between the components in the room.

Pedestrian. A person on foot.

Pedestrian Barricade. An obstruction, obstacle, or barrier set up to check or control crowds of people and foot traffic.

Pedestrian Grade Separation. A structure erected over or under an obstacle such as a freeway, roadway, street, railroad, or stream, and intended primarily for pedestrian use.

Pedestrian Ramp. A sloping path of travel intended for pedestrian traffic and as differentiated from a curb ramp.

Pedestrian Spatial Requirements. Normally 13 square feet are required per person for comfortable motion while walking.

Pedestrian Way. A route by which a pedestrian may pass.

Pediment. The triangular front part of a building, surmounting the columns, of a classical Grecian style.

Peel. To convert a wood log into veneer by rotary cutting.

Peeler. In gypsumboard, the clean separation of surfacing papers from the core; may be due to a variety of causes, most frequently, however, from surface calcination during the drying process or insufficient binder.

Peeling. 1. Detachment of a paint film in relatively large pieces; paint applied to a damp or greasy, surface usually peels; sometimes it is due to moisture back of the painted surface. 2. A process in which thin flakes of matrix or mortar are broken away from concrete surface, caused by adherence of surface mortar to forms as forms are removed. 3. See Shivering.

Peen. The working face of a hammer opposite the flat face, usually shaped for bending, shaping, or cutting the material being struck with the peen; see Cross Peen, Ball Peen, or Straight Peen Hammers.

Peer Review. A quality improvement system consisting of having an architect's or engineer's professional work and office procedures reviewed by a committee of peers.

Pegboard. A board with holes in which pegs or fittings may be inserted to hang tools or objects.

Pegged Flooring. Hardwood flooring with hardwood plugs set to hide recessed screws or to simulate screw covers.

PEI. Porcelain Enamel Institute.

Pein. Variant of peen.

PEL. Permissible Exposure Limits; standards set by OSHA.

Peltier Effect. When direct current is passed through two adjacent metals one junction will become cooler and the other will become warmer; this principle is the basis of thermoelectric refrigeration.

Penal Sum. The face amount of a surety bond; this is the maximum amount of the surety's liability.

Penalty. A sum to be forfeited if a condition is not met.

Penalty and Bonus Clause. A contract provision that the contractor will pay a penalty for late completion and receive a bonus for timely or early completion.

Penalty Clause. A clause in a construction contract by which a contractor is assessed with a monetary penalty, usually on a daily basis, for delay in the completion of a project.

Pencil Rods. Smooth mild steel reinforcing rods of 3/16, 1/4, or 3/8 inch diameter.

Pendentive. One of the concave triangular segments that form the transition from the supporting columns to a dome above.

Penetrating Stain. Stain made by dissolving oil-soluble dyes in oil or alcohol.

Penetrating Stain Wax. Wood finish which produces color of penetrating stain with luster of wax.

Penetration. The consistency of a bituminous material expressed as the distance in tenths of a millimeter (0.1mm) that a standard needle penetrates vertically a sample of the material under specified conditions of loading, time, and temperature.

Penetration Grading. Of asphalt cements, a classification system based on penetration in 0.1mm at 25° C (77° F); in this system, there are five standard paving grades, 40-50, 60-70, 85-100, 120-150 and 200-300.

Penny. As applied to nails, it originally indicated the price per hundred; the term now serves as a measure of nail length and is abbreviated by the letter d.

Pentaerythritol Resins. Resin made by reacting pentaerythritol, a high alcohol, with rosin.

Pentagon. A five sided plane figure.

Penthouse. A top story of a building that is smaller than the story below it.

Penthouse Louver. A louvered wall around the mechanical penthouse area of a structure, with fixed or movable flaps; this protects it from the elements and provides a visual screen around equipment.

People Competence. The aptitude for and fulfillment of necessary management of personnel to complete projects in a professional and timely manner.

Percentage. A part of a whole expressed in hundredths.

Percentage Humidity. The degree of possible saturation of air with water vapor multiplied by 100.

Percentage Lease. A lease in which the rental consideration is an agreed percentage of sales, usually with an expressed minimum rent.

Perched Water Table. Underground water lying over dry soil and sealed from it by an impervious layer.

Percipient Witness. A person called to testify on account of personally observing an event or being personally involved.

Percolate. 1. Seep or ooze. 2. To cause a liquid to pass through a permeable substance.

Per Diem. An allowance or payment for each day.

Perforated PVC Pipe. Plastic pipe 4 inches in diameter with one or more rows of uniform holes along the length. Buried in the ground alongside building foundations or structures, to aid in drainage of groundwater and moisture.

Perforated Strap. Thin metal strips in rolls with punched holes used to hang or keep plumbing pipes in place.

Perforated Wall. Wall with relatively small openings as for ornamentation.

Performance Bond. A bond, secured by the general contractor, which guarantees that the contract will be performed; an undertaking by a surety that a contractor will perform a contract.

Performance Code. A building code that prescribes the objectives sought rather than the specific methods and materials that must be used; compare with Specification Code.

Performance Indicator. The bottom line showing profit or loss on an income and expense statement is the important indicator of the performance of a business.

Performance Specification. A building specification that prescribes the objectives sought rather than the specific methods and materials that must be used.

Pergola. An arbor or covered walk formed of growing plants trained over trellis work supported on parallel rows of columns.

Perilla Oil. Drying oil obtained from seeds of brush called Perilla Ocymoide, grown largely in China and Japan.

Perimeter. The length of the circumference or outline of a figure.

Perimeter Heating. A method of installing central heating systems so that the registers are placed on the outside walls under windows.

Perimeter Overflow System. A continuous channel formed into the sidewall entirely around the perimeter of the pool, unless interrupted by steps, into which surface pool water is continuously drawn during normal operation to provide a skimming action.

Perimeter Relief. 1. Construction detail which allows for building movement. 2. Gasketing materials which relieve stresses at the intersections of wall and ceiling surfaces.

Perineal Bath. A small shallow bathtub in which one bathes in a sitting position; a Sitz bath.

Period. The time for a wave crest to traverse a distance equal to one wave length or the time for two successive wave crests to pass a fixed point.

Periodic. Periodic duty is a type of intermittent duty in which the electrical load conditions are regularly recurrent.

Perjury. Wilfully lying while under oath.

Perlite. Expanded siliceous volcanic rock, expanded by heat, used as a lightweight aggregate in concrete and plaster, and as an insulating fill.

Perlite Insulation. An insulation made from volcanic glass, expanded by heat.

Perlite Roof Insulation. Insulation for a roof system made from volcanic glass, expanded by heat.

Perm. A measure of water vapor movement through a material, such as a vapor barrier; one perm equals one grain of vapor transmission per square foot, per hour, for each inch of mercury difference in vapor pressure.

Permafrost. Subsoil which remains frozen throughout the year, as in the polar regions.

Permanence. The property of a plastic which describes its resistance to appreciable changes in characteristics with time and environment.

Permanent Loan. A long term real estate loan that replaces the construction loan (Interim Loan) upon completion of the construction; also called Takeout Loan.

Permanent Magnet. 1. Material which has its molecules aligned and has its own magnetic field. 2. A bar of metal which has been permanently magnetized.

Permeability. The property of a material to permit a fluid (or gas) to pass through it; in construction, commonly refers to water vapor permeability of a sheet material or assembly and is defined as Water Vapor Permeance per unit thickness. Metric unit of measurement, metric perms per centimeter of thickness. See Water Vapor Transmission, Perm, Permeance (ASTM E 96).

Permeable. Capable of being penetrated; having pores or openings that permit liquids or gases to pass through.

Permeance. A material's resistance to water vapor transmission; the ratio of the rate of water vapor transmission through a material or assembly between its two parallel surfaces to the vapor pressure differential between the surfaces; see Water Vapor Transmission (WVT).

Permutation. An ordered arrangement or grouping of a set of numbers or things; any one of the range of possible groupings.

Perpend Bond. In masonry, a header brick or large stone extending through a wall so that one end appears on each side of the wall and acts as a binder.

Perpendicular. Standing at right angles to the plane of the horizon; vertical.

Perpends. The vertical joints in the face of a wall with all joints directly over one another.

Perpetual. Lasting forever.

Personal Air Samples. An air sample taken with a sampling pump directly attached to the worker with the collecting filter and cassette placed in the worker's breathing zone; these samples are required by the OSHA asbestos standards and the EPA Worker Protection Rule.

Personal Ownership. The ownership of the business by one person; sole ownership or sole proprietor.

Personal Property. Any property that is not real property; also called Chattel.

Personnel. The body of people employed in a business or on a project.

Personnel Lift. An elevator for use by persons at a job site, a building, or structure.

Perspective. A drawing on a plane surface that represents an object as it appears to the eye; it cannot be scaled as the lines are foreshortened.

PERT. Program Evaluation Review Technique.

Pest Control. The act or process of the placement of devices or spraying of chemicals or powders to control the spread of insects and pests.

Petcock. A small valve for draining or letting out air.

Petty Cash Fund. A fund of cash that is established to make minor disbursements and avoid writing checks for small amounts.

Pew. Bench-like seating in a church.

PH. Measurement of the free hydrogen ion concentration in an aqueous solution.

pH Value. Measure of acidity or alkalinity; pH 7 is neutral; the pH values of acids are less than 7 and of alkalis (bases) greater than 7.

Pharmacy. A room or building devoted to the sale and dispensing of drugs, medical supplies, and related goods.

Phase. Distinct functional operation during a cycle.

Phased Application. The system of applying the felt plies of a built-up roofing membrane in two or more steps, separated by a delay normally of at least one day.

Phenol. A caustic poisonous derivative of benzene present in coal tar and wood tar used in dilute form as an antiseptic and disinfectant.

Phenol-Adelhyde Resins. Resins produced from phenols and formaldehyde.

Phenolic Resins. Resin based essentially on reaction between phenol and formaldehyde.

Phenolic-Resin Primer Sealer. Finish well suited for fir and other softwoods, which penetrates into pores of wood, dries and equalizes density of hard and soft grains.

Phenol-Red. A dye which is yellow at a pH of 6.8 and turns a progressively deeper red color as the pH increases to 8.4. This is the most commonly used test reagent for pH in swimming pools.

Phial. The sensing element on a thermostatic expansion valve.

Phillips Screwdriver. A cross-tipped screwdriver for installing screws.

Phloem. In a tree, the tissues of the inner bark, characterized by the presence of sieve tubes and serving for the transport of elaborate foodstuffs.

Phon. A unit of loudness for an average listener of a sound; equal to the sound level being zero at faintest, and 1000 at loudest.

Phosphatize. Form a thin inert phosphate coating on surface usually by treatment with H_3PO_4, phosphoric acid.

Phosphorescent Paint. Luminous paint which emits light after the white light has been turned off; no phosphorus is used.

Photoelectricity. Physical action wherein an electrical flow is generated by light waves.

Photoelectric Sensor. A device that responds to light and transmits a resulting impulse.

Photo Equipment. Materials and devices used for photography and the processing of photographs.

Photo Processor. A piece of equipment used to develop photographs from negatives.

Photogrammetry. The science of reliable measuring of aerial photographs to produce topographic maps.

Photographs. Pictures taken before a job commences to provide an accurate representation of what the site was like before construction.

Photometrics. The science of measuring the intensity of light.

Photo-Oxidation. Oxidation caused by solar rays.

Photovoltaic. A process of conversion of sunlight into electricity.

Photovoltaic Cells. Semiconductor devices that convert solar energy into electrical power.

Phthalic Anhydride. A white crystalline material used in making synthetic resins.

Phthalic Resins. A particular group of film formers; alkyd resins.

Phthalocyanine Blue. Organic blue pigment developed synthetically; outstanding in fade resistance.

Phthalocyanine Green. Complex copper compound pigment with bluish-green cast.

Physical Inventory. A listing of merchandise on hand, determined by actual count, weight, or measurement and the pricing or value thereof.

Physical Resources. The building activity, management and labor personnel and expertise in the operating area of the business as well as scrap materials and equipment held for resale and other assets not listed on a balance sheet.

Physical Therapy. The treatment of an injury or illness by physical and mechanical means, such as massage or heat.

Pi. Designated by the Greek letter P; the symbol of the ratio of the circumference of a circle to its diameter; the value of P is approximately 3.14159265; P = C/D

Piano Hinge. A Continuous Hinge.

Pick. A heavy hand implement with two pointed ends used in digging and loosening earth.

Pick and Dip. A method of laying brick whereby the bricklayer simultaneously picks up a brick with one hand and, with the other, enough mortar on a trowel to lay the brick; also called the Eastern or New England method.

Pickling. A dipping process for cleaning steel and other metals; the pickling agent is usually an acid.

Pick Tong. A blacksmith's tool used to handle hot metal.

Pick Up Sags. When a too-heavy coating of paint has been applied and starts to sag or run down the surface, the painter brushes up through the sagging paint to level it off.

Pickup Truck. A light truck having an open body with low sides and tailboard.

Picture Framing. A rectangular pattern of ridges in a membrane over insulation or deck joints.

Picture Mold. See Picture Mold.

Picture Molding. A molding shaped to form a support for picture hooks, often placed at some distance from the ceiling upon the wall to form the lower edge of the frieze.

Picture Window. A large, often fixed, window, usually of plate or insulating glass, designed to frame an exterior view.

Piece Dyeing. Immersing an entire carpet in a dye bath to produce single- or multi-color pattern effects; see also Resist Printing.

Pier. 1. Timber, concrete, or masonry supports for girders, posts, or arches. 2. An isolated column or mass of masonry units 3. Intermediate supports for a bridge span. 4. Structure extending outward from shore into water used as a dock for ships.

Piezometer. An instrument for measuring pressure or compressibility.

Pigment. 1. The coloring matter in paint or other materials. 2. Material in the form of fine powders insoluble in oils, varnishes, lacquers, thinners and the like; used to impart color, opacity, certain consistency characteristics, and other effects.

Pigment Grind. Dispersion of pigment in a liquid vehicle.

Pigment Oil Stain. Consists of finely ground insoluble color pigments such as used in paints, in solution with linseed oil, varnish, mineral spirits, etc. according to formula being used; also called Wiping Stain.

Pigment Volume Concentration (PVC). Percent by volume occupied by pigment in dried film.

Pilaster. A projecting square column forming part of a wall.

Pilaster Block. A concrete masonry unit that allows construction of a pilaster in a concrete block wall.

Pile. 1. Long steel, wood, or concrete member penetrating deep into the soil to support grade-beam foundation walls or columns. 2. The raised yarn tufts of woven, tufted and knitted carpets which provide the wearing surface and desired color, design or texture; in flocked carpets, the upstanding, non-woven fibers.

Pile Butt. Thelarge end of a pile; the small end is called the Tip.

Pile Cap. A thick slab of reinforced concrete poured across the top of a pile group to cause the group to act as a unit in supporting a column.

Pile Cap, Concrete. A concrete footing resting on a group of piles.

Pile Cap Formwork. Formwork for a concrete pile cap.

Pile Cap Reinforcing. Steel reinforcing bars in a concrete pile cap.

Pile Crushing. Of carpet, the bending of pile due to foot traffic or the pressure of furniture.

Pile Group. Several driven or poured piles in a cluster and tied together at their tops by a single poured concrete pile cap.

Pile Height. Of carpet, the height of pile measured from the top surface of the backing to the top surface of the pile; also referred to as Pile Wire Height.

Pile Setting. Of carpet, brushing after shampooing to restore the damp pile to its original height.

Pile Spall. A chip or piece broken from a pile by a blow from the driving hammer or by action of the elements.

Pile Testing. The act or process of measuring the resistance of a driven test pile to a pre-determined design load.

Pile Tip. Thesmall end of a pile; the large end is called the Butt.

Pile Warp. Lengthwise pile yarns in Wilton carpets which form part of the backing.

Pile Wire Height. See Pile Height.

Pile Yarn. Of carpet, the yarn used to form the loops or tufts of a pile fabric.

Pile Yarn Density. Of carpet, the weight of pile yarn per unit of volume in carpet, usually stated in ounces. per cubic. yard.

Pillar. A column or post supporting a roof or for ornamentation.

Pilling. Appearance defect associated with some staple fibers where balls of tangled fibers are formed on the carpet surface which are not removed readily by vacuuming or foot traffic; pills can be removed by periodic clipping.

Pilot Circuit. Secondary circuit used to control a main circuit or a device in the main circuit.

Pilot Hole. A small hole drilled into a piece of wood to prevent it from splitting when it receives a screw or nail.

Pilot Light. 1. A small light to indicate when a switch or other electrical device is on or in the on position. 2. A relatively small flame which may be automatic or may burn continuously; its purpose is to ignite the main supply of gas when a gas-fired heating or cooking unit is turned on.

Pin, Clevis. See Clevis Pin.

Pinhole. 1. Any small hole. 2. A small perforation in the gypsum board paper or paper joint tape. 3. A small hole appearing in a cast when the water-stucco ratio has not been accurately measured; excess water causes pinholes.

Pinion Gear. 1. A small gear in a set of gear wheels. 2. The smaller gear of two or the smallest of three or more.

Pin Joint. A hinge in a structure; a structural joint that will not transmit moment.

Pin Knot. A knot that is not more than 1/2-inch in diameter.

Pinnacle. A small ornamental turret usually ending in a pyramid or cone, crowning a buttress or roof.

Pinstripes. Fine stripes.

Pin Tumblers. Small sliding pins in a lock cylinder working against coil springs which prevent the cylinder plug from rotating until the pins are raised to alignment by bitting of the proper key.

Pipe. 1. A long tube or hollow body for conducting a liquid, gas, or finely divided solid. 2. A structural column or strut.

Pipe, ABS. A plastic pipe made in various diameters; used for stacks and main drains in plumbing systems.

Pipe, Aluminum. A pipe for liquid or gas.

Pipe Bollard. Short pipe length, placed vertically in the ground and filled with concrete to prevent vehicular access or to protect property from damage by vehicular encroachment.

Pipe, Clay. Pipe used for drainage systems and sanitary sewers made of earthenware and glazed to eliminate porosity.

Pipe Cleanout. See Cleanout Plug.

Pipe, Copper. Rigid pipe used for exterior and interior water systems, it is light and durable, resists moisture, and can be used with either mechanical or soldered connections.

Pipe, CPVC. Rigid plastic pipe used for hot and cold water supply lines.

Pipe, Fiberglass. A pipe for liquid or gas, fabricated from layers of glass fibers and resins.

Pipe Flange. Projecting ring, ridge or collar placed on pipe to strengthen, prevent sliding, or to accommodate attachments.

Pipe, Galvanized. Zinc coated steel pipe.

Pipe, Glass. Glass and glass-lined pipe used in process piping.

Pipe Handrail. A metal pipe used as a handrail.

Pipe Insulation. Insulation that covers pipes to help in the reduction of heat loss or gain.

Pipe Jacking. Forcing pipe through the ground in a tunnel created by the pipe itself; the pipe is generally jacked horizontally in short lengths.

Pipe Joint Compound. Putty-like material used to seal threaded pipe joints.

Pipe, No-Hub. Pipe manufactured in cast iron, which is fabricated without hubs for coupling.

Pipe Painting. The act or process of painting piping to prevent rust and corrosion and also provide an identifying system in buildings and process piping plants.

Pipe, Polyethylene. Pipe manufactured from a thermoplastic compound.

Pipe, Polypropylene. A tough plastic pipe with resistance to chemicals and heat.

Pipe, PVC. Polyvinyl chloride pipe used mainly for drain lines, particularly resistant to chemicals.

Pipe Railing. A metal railing made of pipe.

Pipe Sleeve. Cylindrical insert cast into concrete wall or floor to provide for later passage or anchorage of pipe.

Pipe, Stainless. Pipe or tubing constructed of stainless steel which has a high resistance to corrosion.

Pipe, Structural. Pipe used in a structure to transfer imposed loads to the ground.

Pipe Wrench. A wrench with serrated jaws to grip pipe and turn it in one direction only.

Piping. Any system of pipes in a building.

PIR Detector. Passive Infra-Red detector. A part of a burglar alarm system.

Piscina. A basin with a drain, usually to a soak pit, near the altar in a church for disposing of water from liturgical ablutions.

Piston. Close-fitting part or plug which moves up and down in a cylinder.

Piston Displacement. Volume displaced by piston as it travels the full length of its stroke; volume obtained by multiplying area of cylinder bore by length of piston stroke.

Pit. A hole, shaft, or cavity in the earth.

Pitch. 1. Resin obtained from various conifers. 2. A black or dark viscous substance obtained as a residue in distilling tar or oil from bones; it also occurs in natural form as asphalt. 3. The slope of a stairway, in degrees. 4. The slope of a roof. 5. Degree of slope or grade given a horizontal run of pipe. 6. The spacing of rivets or bolts. 7. The frequency of sound vibrations. 8. In carpet, the number of pile ends per inch of width; actually, in practical floor covering specifications, it is taken as the number of pile ends per unit of standard 27-inch width; terms of pitch used commonly in the industry are 180, 189, 192, 216 and 256.

Pitch Board. See Bevel Board.

Pitch Pine. Any of the pines that yield pitch; the heartwood is brownish red and resinous; the sapwood is thick and light yellow; used for lumber, fuel, and pulpwood.

Pitch Pocket. 1. An opening between growth rings of a tree which usually contains resin, bark, or both. 2. In roof construction, a flanged metal container placed around a roof penetration at roof level to receive hot bitumen or caulking and provide a roof seal; commonly found at columns or plumbing stacks.

Pitch Streak. A well-defined accumulation of pitch in a more or less regular streak in the wood of certain conifers.

Pith. The small, soft core occurring near the center of a tree trunk, branch, twig, or log.

Pitot Tube. Tube used to measure air velocities.

Pits. See Pops.

Pitting. Formation of small, usually shallow depressions or cavities in the surface of a material.

Pivot. A shaft, axis, or pin on which something turns or oscillates.

Placement. Process of placing and consolidating concrete; also called Pour.

Plain Bar. Steel reinforcing bar without deformations.

Plain Concrete. Concrete that is either unreinforced or contains less reinforcement than the minimum amount specified in the code for reinforced concrete.

Plain Reinforcement. See Plain Bar.

Plainsawed Lumber. 1. Flat-grained lumber. 2. Lumber sawed regardless of the grain, the log is simply squared and sawed to the desired thickness; also called Slash Sawed or Bastard Sawed.

Plan. A drawing or diagram made by projection on a horizontal plane; a floor plan is a horizontal section through a building.

Planar. Of or in a plane or level surface, two dimensional

Plancher. See Plancier.

Plancier. 1. A soffit under a cornice or any projecting member; also called Plancher. 2. A plank floor or platform.

Plane. 1. A level surface; a flat surface on which a straight line joining any two points on it would wholly lie. 2. A tool for smoothing or shaping a wood surface.

Plane of Weakness. The plane along which a structure under stress will tend to fracture; may exist because of the nature of the structure and its loading, by accident or by design.

Planing Mill Products. Products worked to pattern, such as flooring, ceiling, and siding.

Plank. A wide piece of sawed timber, usually 1-1/2 to 4-1/2 inches thick and 6 inches or more wide.

Plank Floor. See Random Width Flooring.

Planned Unit Development. A cluster development that includes more land uses in the built up area (industrial, commercial) than the residential cluster system.

Planning. The process of carrying out plans for development of land or buildings.

Plant. 1. To set or place plants or seeds in the ground. 2. The trees, shrubs, or seeds to be planted. 3. The land, buildings, and equipment used for processing or manufacturing.

Plant Bed Preparation. The mixing of earth with fertilizer and other soil enriching products, to create an area for the future introduction of vegetation.

Planter. A container in which plants are grown.

Planting. 1. A plantation. 2. A method of placing spools of different colors of surface yarn in frames back of Jacquard Wilton looms so that more colors will appear in the design than are supplied in the full solid colors used; these extra planted colors are usually arranged in groups of each shade, to give added interest to the pattern.

Plantroom. Operational pool equipment location.

Plant Screens. In a concrete batching plant, screens located between the dryer and hot bins which separate the heated aggregates into the proper hot bin sizes.

Plaque. 1. A commemorative or identifying inscribed tablet. 2. A localized abnormal patch on a surface.

Plaster. A cementitious material, usually based on gypsum or portland cement, which is applied to lath or masonry in paste form, to harden into a finished surface; a mixture of lime, hair, and sand, or of lime, cement, and sand, used to cover exterior or interior wall surfaces; from Greek, emplastron, to daub on; Latin, emplastrum; French, platre; old English, plaister.

Plaster Accessory. Hardware or tool needed for plaster work.

Plaster Base. The lath or backing to which plaster is applied.

Plaster Bead. Built-in edging usually metal, to strengthen a plaster angle.

Plasterboard. A board used in large sheets as a backing or as a substitute for plaster in walls and consisting of fiberboard, paper, or felt, bonded to a hardened gypsum plaster core; see Gypsumboard.

Plaster Bond. The state of adherence between plaster coats or between plaster and a plaster base, produced by adhesive or mechanical interlock of plaster with base or special supplementary materials.

Plaster Fascia. The exposed vertical face of a wall cornice in an interior location, finished in plaster.

Plastering. The work produced by plasterers.

Plastering Machine. A mechanical device by which plaster mortar is conveyed through a flexible hose to deposit the plaster in place; also known as a plaster pump or plastering gun; distinct from Gunite machines in which the plaster or concrete is conveyed, dry, through the flexible hose and hydrated at the nozzle.

Plaster, Masonry. See Masonry Plaster.

Plaster of Paris. A fine white gypsum plaster used for casting and molding; pure calcined gypsum.

Plaster Patch. Plaster mixed or pre-mixed to fill cracks and damaged areas.

Plaster Ring. A guide with a metal collar attached to a base to apply plaster to a certain thickness or provide a fastener for trim.

Plaster Screed. 1. A strip of wood or metal to regulate the thickness of plaster and furnish an edge trim. 2. A spot or strip of plaster to regulate plaster thickness.

Plaster Work. The finished product of plasterers.

Plastic. 1. Capable of being molded; pliant; supple. 2. Any of a number of synthetic polymeric substances that can be given any required shape. 3. In stress analysis, refers to stress/strain behavior beyond the elastic range; plastic deformation usually implies some permanent shape change. 4. A condition of freshly mixed concrete, mortar, or cement paste indicating that it is workable and readily remoldable, is cohesive and has an ample content of fines and cement but is not over wet.

Plastic Cement. 1. A plastic mixture of bitumen and asbestos reinforcing fibers with a solvent; see Flashing Cement. 2. Portland cement to which small amounts of plastizing agents, not more than 12% by volume, have been added at the mill.

Plastic Coated Conduit. A type of conduit for electrical wiring that is used around moist areas and highly corrosive fumes.

Plastic Consistency. Condition in which concrete, mortar, or cement paste will sustain deformation continuously in any direction without rupture.

Plastic Design. Method of structural analysis of continuous steel structures (beams and frames) based on calculating the loading which will cause collapse of the system.

Plastic Hinge. The point at which a structural member such as a beam or part of a frame is bent so that a section is stressed beyond the elastic range.

Plasticity. 1. Property of a material to deform under load and to retain the deformation after the load is removed. 2. A material's ability to be shaped and worked. 3. That property of plaster mortar that permits continuous and permanent deformation in any direction. 4. A plastic material is distinct from a fluid material in that it requires a measurable force (yield value) to start flow; the property exists in varying degrees in different materials and in plaster mortar is sometimes regarded as an index of working characteristics.

Plasticizer. 1. A substance added to plastics, resins, and rubbers to impart improved properties such as flexibility, workability, or stretchability. 2. A material that increases the workability or consistency of a concrete mixture, mortar or cement paste. 3. A product used to increase the flow and/or workability of plaster.

Plasticizing Agent. Plasticizer.

Plasticizing Agent. A product used to increase the flow and/or workability of plaster.

Plasticizing Wood. Softening wood by hot water, steam, or chemical treatment to make it easier to mold.

Plastic Laminate Backing Sheet. Sheet material placed on concealed side of the panel material similar to sheet material on exposed side of panel to balance the construction of the panel and give dimensional stability by minimizing moisture absorption.

Plastic Laminate Casework. Assembled plastic laminate cabinets.

Plastic Laminate Countertop. Countertop with plastic laminate finish covering substrate, usually plywood or particleboard.

Plastic Laminate. Sheet material manufactured of multiple layers of paper with top layer of plastic usually 1/16 inch (1.59 mm) thick with decorative finish; may be used in flat sheets or heat formed, bent, and adhered to single curved base material; commonly referred to by the brand name of Formica.

Plastic Pipe. Pipe manufactured from hard plastic to resist corrosion and rust.

Plastic Sheet. Plastic in which thickness is very small in relation to length or width.

Plastic Skylight. A transparent or translucent plastic molded unit that is set in a frame and mounted on a roof for use as a skylight.

Plastic Soil. A soil that can be rolled into 1/8" diameter strings without crumbling; a soft rubbery soil.

Plastisol. Film former containing resin and plasticizer with no solvents.

Plat. 1. A small piece or ground; a plot of land. 2. A drawing or map showing a plot of land.

Plate. The horizontal framing member at the top or base of wood stud wall framing.

Plate, Armor. See Armor Plate.

Plate Cut. The cut at the bottom end of a rafter to allow it to fit upon the plate; the cut in a rafter which rests upon the plate; also called the Seat Cut.

Plate, Duplex. See Duplex Plate.

Plate Girder. A large beam made up of steel plates, sometimes in combination with steel angles, welded, bolted or riveted together.

Plate Glass. Glass of high optical quality produced by grinding and polishing both faces of a glass sheet.

Plate Mirror. Thick mirror glass manufactured to a high-quality standard.

Plate Tectonics. The theory and study of plate formation, movement, and interaction; the theory which explains seismicity in terms of plate movement.

Plate, Switch. See Switch Plate.

Plate, Toe. See Toe Plate.

Plate, Ultrasonic Examined. Steel plate ultrasonically examined for structural defects.

Platform. Horizontal landing in stair either at the end of a flight or between flights, either at floor level or between floors.

Platform Frame. A wooden building frame composed of closely spaced members nominally 2 inches in thickness, in which the wall members do not run past the floor framing members; typical wood stud wall framing in which the studs are one level in height and the floor framing above rests on the top plates of the wall below; the most common type of framing used in house construction; also called Western Frame; compare Balloon Frame.

Plating. A thin coating of metal deposited on a surface, usually by electrolysis.

Player Bench. A long seat, with or without seat back, on which athletes can sit when they are not playing.

Playground Equipment. Devices such as slides and swings, that children play on or upon in a park or playground.

Pleadings. Formal written documents, filed with a court, accusing a party of wrongdoing, or defending a party against such an accusation.

Plenum. 1. An enclosed chamber such as the space between a suspended finished ceiling and the floor above. 2. Chamber attached directly to a furnace which receives heated air; from this largest chamber, ducts carry the air to the various registers.

Plenum Chamber. Chamber or container for moving air or other gas under a slight positive pressure.

Plenum, Ceiling. See Ceiling Plenum.

Plexiglass. A plastic resilient material comparable to glass in use, usually manufactured in sheets.

Pliers. A hand tool used to grip objects.

Plinth. 1. The lower square slab at the base of a column. 2. The block of wood at the corner of a doorway where the baseboard and architrave meet.

Plot Plan. A drawing showing a plot of land with all its existing physical features and proposed improvements; a ground plan.

Plough. See Plow.

Plow. To cut a groove running in the same direction as the grain of the wood; also spelled Plough.

Plug. A pipe fitting with external threads and squared head that is used for closing the opening in another fitting.

Plug, Conduit. See Conduit Plug.

Plug, Grounding. See Grounding Plug.

Plug, Lock. See Lock Plug.

Plug Valve. See Core Cock.

Plug Weld. A circular weld made by either arc or gas welding through one member of a lap or tee joint joining that member to the other; the weld may or may not be made through a hole in the first member; if a hole is used, the walls may or may not be parallel and the hole may be partially or completely filled with weld metal.

Plumb. Vertical, or perfectly straight up-and-down; at right angles to horizontal.

Plumb Bob. The metal weight on the bottom of a plumb line.

Plumb Bond. In brick laying, an all stretcher bond with all joints directly over one another.

Plumb Cut. Any cut made in a vertical plane; the vertical cut at the top end of a rafter.

Plumber. A craft worker skilled in the installation, repair, and maintenance of water and waste systems in buildings.

Plumber's Friend. A tool consisting of a large rubber like cup and handle, it is used under water to force blockage through sewer lines.

Plumber's Snake. See Snake.

Plumber's Furnace. A heating source used to melt lead, heat soldering irons, or melt solder.

Plumber's Soil. A mixture of glue and lampblack used in lead work to prevent lead from sticking to selected metal parts of lead pipe and fittings.

Plumbing. 1. The act or process of installing in a building or structure the pipes, fixtures, or other apparatus for supplying potable water and removing liquid and water-borne wastes. 2. The installed fixtures and piping of a building or structure. 3. The gas piping system. 4. The storm sewer system. 5. Heating and air conditioning piping.

Plumbing Access Door. A door in a floor, wall, or ceiling giving access to plumbing fittings, drains, or valves.

Plumbing Fixture. Plumbing equipment, usually installed last, such as sinks, water closets, bidets, and shower/bath units; devices which receive water and discharge it and/or water-borne waste into the DWV system.

Plumb Rule. A narrow board having a plumb line and bob on one end, or more commonly having a bubble in a tube; it is used for establishing vertical lines and horizontal lines.

Plunge. A swimming pool.

Plunger. See Plumber's Friend.

Plunge Router. A router in which the cutting bit enters the work from the surface of the board.

Plush. Of carpet, a smooth-face cut pile surface that does not show any yarn texture.

Plus Pressure. See Positive Pressure, 2.

Ply. 1. A layer or thickness, as of building or roofing paper. 2. A layer of wood in plywood. 3. A layer of felt in a built-up roofing membrane; a four-ply membrane would have at least four plies of felt at any vertical cross section cut through the membrane; the exposure of any felt may be computed by dividing the felt width minus 2 inches by the number of plies, thus, the exposed surface of a 36 inch wide felt in a four-ply built-up roof membrane should be 8-1/2 inches. 4. A layer or thickness of yarns used in carpet; if the pile yarn is described as 4-ply, it means that each tuft is made of 4 yarns spun together.

Ply Sheet. A glass fiber felt sheet coated on both sides with asphalt.

Plywood. A wood product made by bonding together layers of veneer or a combination of veneer layers and a lumber core; the layers are joined with an adhesive; adjoining plies are usually laid with grains at right angles to each other, and almost always an odd number of plies are used.

Plywood Countertop. Plywood panel used as the substrate for the exposed finish surface of a countertop.

Plywood Diaphragm. Plywood sheathing on floors, roofs, or walls which provides shear strength to resist wind and earthquake loads.

Plywood, Finish. See Finish Plywood.

Plywood Roof Sheathing. Plywood sheets secured to exterior side of roof rafters or trusses used to create rigidity in building superstructure and serve as base to receive roofing.

Plywood Shear Wall. Wall constructed of studs and plywood sheathing that in its own plane resists shear forces resulting from applied wind, earthquake or other transverse loads or provides frame stability.

Plywood Sheathing. Wall, roof, or floor sheathing of plywood.

Plywood Shelving. Horizontal mounted plywood surfaces upon which objects may be stored, supported, or displayed. OBV

Plywood Siding. Plywood sheeting material forming the exterior surface of outside walls of frame buildings.

Plywood Soffit. Plywood finish material installed horizontally covering the underside of an assembly such as roof eaves.

Plywood, Structural. A grade of plywood that has been tested and graded for its structural characteristics.

Plywood Subfloor. Plywood sheets secured to the top side of floor joists.

Plywood Subfloor-Underlayment. Plywood sheets secured to top side of floor joists used to create rigidity in building superstructure and also to provide smooth and even surface to receive finish floor covering.

Plywood Underlayment. Flat sheet material of plywood placed over subflooring to provide smooth and even surface to receive finish floor covering.

Plywood Wall Sheathing. Plywood material secured to exterior side of exterior wall studs used to create rigidity in building superstructure and serve as base to receive siding.

PM. Preventive Maintenance.

PMI. Private Mortgage Insurance.

Pneumatically Driven Fastener. Driven pin or threaded stud which is driven into material by use of compressed air.

Pneumatic Concrete. Concrete that is delivered by equipment powered by compressed air.

Pneumatic Hoist. A hoisting system powered by compressed air.

Pneumatics. Study of compressible air and gases, their properties and reaction in containment.

Pneumatic System. Mechanical devices powered by compressed air.

Pneumatic Tool. A tool powered by compressed air.

Pneumatic Tube System. An arrangement of tubes within a building or building complex for sending message capsules by compressed air.

POA Valve. See Pressure-Operated Altitude Valve.

Poché. A drafting technique in which parts of drawings are shaded in for increased readability, such as the walls on floor plans or shadows on elevations.

Pocket, Pitch. See Pitch Pocket.

Pock Marks. 1. Pits; craters. 2. In gypsum board, surface imperfections appearing as a multitude of small depressions; often caused by dirt or small gravel which indented the paper surface.

Point. 1. A geometric element without dimension. 2. A sharpened tip as of a tool. 3. In carpet, one tuft of pile.

Pointing. 1. Troweling mortar into a joint after masonry units are laid. 2. The process of inserting mortar into the joints or brickwork or any other masonry units to fill open pockets or voids left when the work was originally done.

Pointing Trowel. A mason's tool used to point brick, stone, and block.

Point of Inflection. The point in a structural member where the bending moment changes from positive to negative.

Points. A fee charged for making a loan, each point amounting to one percent of the loan amount.

Point Weight. See Felt Mill Ream.

Poise. A centimeter-gram-second unit of absolute viscosity, equal to the viscosity of a fluid in which a stress of one dyne per square centimeter is required to maintain a difference of velocity of one centimeter per second between two parallel planes in the fluid that lie in the direction of flow and are separated by a distance of one centimeter.

Polarity. 1. The condition of having two poles with contrary qualities. 2. The positive or negative electrical condition.

Polarized Plug. An electric plug with a configuration of prongs that prevent plugging in improperly.

Polarizing. This is the method of identifying electrical wires by colors. It is to make sure that hot wires will be connected only to hot wires and that neutral wires will run in continuous circuits without interruption back to the ground terminals.

Polder. A low lying area of land that has been reclaimed from the sea.

Pole. 1. One of the two terminals of an electric cell, battery or dynamo. 2. Either extremity of the axis of a sphere. 3. The vertex of an angle coordinate. 4. A point of guidance. 5. A stake.

Pole, Closet. See Closet Pole.

Pole, Electric. See Electric Pole.

Pole-Gun. Spray gun equipped with an extension tube.

Pole, Utility. See Utility Pole.

Police Equipment. Any of assorted devices necessary for law enforcement personnel to carry out their duties.

Police Power. The inherent power of the state to regulate its citizens and property in the interest of general security, morals, health, and welfare.

Police Station. A building housing the police and their activities in a locality.

Polish. 1.To make smooth and shiny by rubbing. 2. To make plaster finish coat smooth and glossy by troweling.

Polished Plate Glass. Glass 1/4-inch or thicker that is made by a process where both sides are ground and polished

Polishing. Said of wall paints where shiny spots or surfaces have resulted from washing or wiping.

Pollution. Contamination of materials or the environment by harmful substances.

Polyamide. Product used in making dripless paint; see thixotropic paint.

Polybutylene Plastics. Plastics based on polymers made with butenas essentially the sole monomer.

Polychlorinated Biphenyl (PCB). Compound used in electrical transformers that is a poisonous environmental pollutant.

Polychromatic. Multi-colored.

Polychrome Finish. Finish obtained by blending together a number of colors.

Polyester Floor. Flooring material that has adhesive properties, high strength, and good chemical resistance, manufactured from a synthetic resin that polymerizes during curing.

Polyethylene. A polymer of ethylene, a thermoplastic used in packaging and insulation; plastic sheeting.

Polyethylene Pipe. Pipe manufactured from a thermoplastic, high-molecular-weight, organic compound.

Polyethylene Vapor Barrier. A sheet form thermoplastic membrane, high molecular weight, organic compound, used as a protective cover to prevent the passage of air or moisture.

Polyethylene Wrap. A sheet form thermoplastic high-molecular-weight organic compound used to protect concrete during curing or as a temporary enclosure for construction operations.

Polygon. A plane figure with 3 or more sides.

Polymer. A compound composed of one or more large molecules that are formed from repeated units of smaller molecules.

Polymeric. Composed of repeating chemical units; all plastics and polymers are polymeric.

Polymerization. A chemical reaction in which two or more small molecules combine to form larger molecules that contain repeating structural units of the original molecules; when two or more monomers are involved, the process is called copolymerization or heteropolymerization.

Polyolefin Plastics. Plastics based on polymers made with an olefin as essentially the sole monomer.

Polyphase Motor. Electrical motor designed to be used with a three or four-phase electrical circuit.

Polypropylene. See Olefin.

Polypropylene Pipe. A tough plastic pipe with resistance to chemicals and heat.

Polypropylene Plastics. Plastics based on polymers made with propylene as essentially the sole monomer.

Polystyrene. A plastic based on a resin made by polymerization of styrene as the sole monomer; polystyrene may contain minor proportions of lubricants, stabilizers, fillers, pigments and dyes; a plastic foam board used for insulation.

Polyurethane. 1. Any polymer containing urethane, used in plastics, adhesives, paints, foams, and rubbers. 2. Various polymers that are used in flexible and rigid foams, elastomers, and resins.

Polyvinyl Acetate (PVA). 1. A synthetic resin used extensively in emulsion (water) paints; produced by the polymerization of vinyl acetate. 2. White glue used in woodworking.

Polyvinyl Acetate Latex (PVA Latex). Latex composed primarily of vinyl acetate.

Polyvinyl Butyral. Plastic film used as an interlayer in laminated glass.

Polyvinyl Chloride. 1. A tough transparent solid polymer, easily colored, used for pipes, fittings, flooring, and a wide variety of other plastic products. 2. A synthetic resin used in solvent type coatings and fluid bed coatings, produced by the polymerization of vinyl chloride; PVC is also used in emulsion (water) paints.

Polyvinyl Chloride Plastics (PVC). Plastics made by combining polyvinyl chloride with clorants, fillers, plasticizers, stabilizers, lubricants, other polymers, and other compounding ingredients; not all of these modifiers are used in pipe compounds.

Polyvinyl Resin Emulsion Glue. White glue; wood adhesive intended for interiors; made from polyvinyl acetates which are thermoplastic and not suited for temperatures over 165° F.

Pond. 1. A small lake. 2. A roof surface which is incompletely drained.

Ponded Roof. Flat roof designed to hold a quantity of water which acts as a cooling device.

Ponding. 1. Water puddles standing on a roof due usually to improper drainage or deck deflection; also called Birdbaths. 2. Water puddles on concrete paving or asphaltic concrete paving due to insufficient sloping or imperfect finishing.

Pool. A body of water; or bath, hydro, hot tub, lido, pond, spa, splasher, and thermae, or covered balneum, bath house, leisure centre, natatorium, poolarium, pool enclosure, pool hall, and pool house.

Pool Depths. The distance between the floor of pool and the maximum operating water level.

Pool Equipment. Implements and devices used for pool maintenance and operation.

Pool Ladder. A ladder used for climbing in and out of a swimming pool.

Pool Light. A light mounted in the wall of a pool for night use.

Pool Liner. Waterproof membrane for pool; usually flexible vinyl.

Pool Plumbing. All chemical, circulation, filter waste discharge piping, deck drainage and water filling system.

Pool Shell. Pool floor and walling structure; also called Pool Tank.

Pool Tank. See Pool Shell.

Pop-Off Valve. A safety valve which opens automatically when pressure exceeds a predetermined limit.

Pop Rivet. Metal fastener for joining sheetmetal pieces, installed by a hand operated compressed air-assisted or spring-loaded gun; unique in that installation may be from one side of the work.

Pops. Ruptures in finished plaster or cement surfaces which may be caused by expansion of improperly slaked particles of lime or by foreign substances; also called Pits.

Pop-Up Drain. The part of a sink drain assembly that is operated by a linkage to open or close the drain.

Pop-Up Head. In a lawn irrigation system, a watering head that retracts when not in use, and becomes flush with the ground.

Porcelain. Ceramic china-like coating applied to steel or cast iron surfaces; when used as a finish for metal fixtures it is called vitreous enamel.

Porcelain Cleat. A ceramic electrical insulator.

Porcelain Enamel. Vitreous Enamel.

Porcelain Mullite. A vitreous ceramic whiteware for technical application in which mullite ($3A_{12}O_{3}2SiO_{2}$) is thc essential crystalline phase.

Porcelain Receptacle. A simple electrical lighting fixture consisting of a bulb holder, with or without a pull chain switch.

Porcelain Sink. A wash basin made from nonporous hard white glazed ceramic coated steel or cast iron.

Porch. 1. An ornamental entrance way to a house. 2. A landing before a door.

Pores. Openings on the surface of a piece of wood; these openings result when vessels in the wood are severed during sawing; see also Vessels.

Porosity. 1. Being permeable to liquids. 2. Degree of integrity or continuity. 3. Gas pockets or voids in metal. 4. The ratio of the volume of voids in the material to the total volume of the material, including the voids, usually expressed as a percentage.

Porous Fill. Soil that allows relatively free passage of water.

Porous Woods. Hardwoods having vessels or pores large enough to be seen readily without magnification.

Porphyry. A hard igneous rock quarried in ancient Egypt, composed of crystals of white or red feldspar in a red matrix.

Portable Bleacher. An uncovered stand of tiered planks that can be moved from one event to another.

Portable Stage. A movable raised platform usually used for entertainment and speaking purposes.

Portal. An elaborate or imposing door or entrance.

Portal Frame. A rigid frame; two columns and a beam attached with moment connections.

Portcullis. A strong heavy grating of iron or iron-bound wood sliding up and down in vertical grooves to block a gateway in a fortress.

Porte Cochere. A roofed structure adjacent to a building entrance covering a driveway, to allow shelter to automobile passengers.

Portico. A roofed colonnade adjoining a building

Portland Blast Furnace Slag Cement. The product obtained by intimately ingrinding or intimately and uniformly blending a mixture of granulated blast furnace slag and portland cement clinker.

Portland Cement. (1) A hydraulic cement produced by pulverizing clinker consisting essentially of hydraulic calcium silicates, and usually containing one or more of the forms of calcium sulfate as an interground addition; the most common type of cement used in construction; the name portland is not a brand name, it refers to a type of cement. (2) Product obtained by pulverizing and mixing limestone with other products to produce cement required for concrete mix.

Portland Cement Grout. Portland cement added to fine aggregates and water and pumped or forced into joints, cracks, and spaces as an adhesive sealer or structural fill.

Portland Cement Mortar. Mortar where the cementing agent is portland cement.

Portland Cement Plaster. A binder of portland cement mixed with plaster and used on exterior surfaces or in damp areas.

Portland Cement-Lime Plaster. Portland cement and lime (either Type S hydrated lime or properly aged lime putty) combined in suitable proportions.

Portland Pozzolan Cement. The product obtained by intimately intergrinding a mixture of portland cement clinker and pozzolan, or an intimate and uniform blend of portland cement and fine pozzolan.

Positioned Weld. A weld made in a joint which has been so placed as to facilitate making the weld.

Positioner. Wire device designed to hold reinforcing steel in designated position in a masonry wall.

Positive Drainage. In roof construction, the drainage condition in which consideration has been made for all loading deflections of the deck and additional roof slope has been provided to ensure complete drainage of the roof area within 24 hours of rainfall precipitation.

Positive Moment. A result of bending moment in a beam in which the upper part is in compression and the lower part is in tension; compare with Negative Moment.

Positive Pressure. 1. Condition that exists when more air is supplied to a space than is exhausted, so the air pressure within that space is greater than that in surrounding areas. 2. A pressure within the sanitary drainage or vent piping system that is greater than atmospheric pressure; also called Plus Pressure.

Post. A timber set on end to support a wall, girder, or other member of the structure.

Postal Accessories. Implements and devices used to weigh, sort, package, price, and handle mail.

Post Cap. 1. A metal connection from a wooden post to a girder. 2. A prefabricated fitting which tops a wooden post and protects it from the elements.

Post, Corner. See Corner Post.

Postformed Plywood. The product formed when flat plywood is reshaped into a curve configuration by steaming or plasticizing agents.

Post Hole. A dug out hole in the ground for the installation of a fence or gate post.

Post Light. A lighting fixture mounted on a post.

Post Line. The line which marks the outside face of a foundation, the location line of fence posts or the line of piers for a deck.

Post-Tensioning. A method by which concrete is compressed after it has been cast by stressing the steel reinforcing; the compressing of the concrete in a structural member by means of tensioning high-strength steel tendons against it after the concrete has cured.

Post, Treated. See Treated Post.

Potable Water. Water that is suitable for drinking.

Potassium Alum. Used as flocculent in sand filter operation.

Potential Energy. A body's ability to do work by virtue of its position relative to others, or by the arrangement of parts.

Potential Relay. Electrical switch which opens on high voltage and closes on low voltage.

Potential, Electrical. See Electrical Potential.

Potentiometer. Instrument for measuring or controlling by sensing small changes in electrical resistance.

Pot Hole. A small steep-sided hole usually with underground drainage.

Pot Life. See Working Life.

Pouch. See Nail Apron.

Pound. 1. A unit of weight equal to 16 ounces avoirdupois or 0.4536 kilograms. 2. Crush or beat with repeated heavy blows.

Poundal. A unit of force equal to the force that would give a free mass of one pound an acceleration of one foot per second per second.

Pounds per Square Inch Absolute (PSIA). Absolute Pressure.

Pounds per Square Inch Gauge (PSIG). Gauge Pressure.

Pour. To place concrete; a continuous increment of concrete casting carried out without interruption.

Pour Coat. The top coating of bitumen on a built up roof.

Poured in Place. Concrete that is poured and cured on site in its final position.

Pour Point. Lowest temperature at which a liquid will pour or flow.

Powder. 1. A substance in the form of fine dry particles. 2. Gunpowder.

Powder-Actuated Fastener. Driven pin or threaded stud made of special heat-treated steel to provide hard ductile fastener which is driven into material by impact of discharge of an explosive cartridge; used to fasten materials to concrete and structural steel.

Powder Room. A guest bathroom, usually containing a toilet and a lavatory; usually located near the front entry hall to be convenient to guests.

Powder Stains. Stains in form of powder which are mixed with solvents to produce wood stains.

Power. 1. A source or means of supplying energy, such as electricity. 2. Time rate at which work is done or energy emitted.

Power-Assisted Door. A door used for human passage with a mechanism that helps to open the door, or relieves the opening resistance of a door, upon the activation of a switch or a continued force applied to the door itself.

Power Driven Fastener. A fastener attached to steel, concrete or masonry by a power charge cartridge or by manual impact.

Power Element. Sensitive element of a temperature-operated control.

Power Factor. 1. The ratio of the true power to the apparent power. 2. Correction coefficient for the changing current and voltage values of a-c power.

Power of Attorney. A legal instrument that authorizes one to act as the agent of another; the agent is called an attorney in fact, but does not have to be a licensed attorney.

Power Stretcher. A carpet layer's tool; an extension-type version of the knee kicker, with larger teeth arranged in a patent head which can be adjusted for depth of bite, used to stretch larger areas of carpet than can be handled by the knee kicker.

Power Tool. An apparatus or device used in construction, powered by electric current.

Power Trowel. See Mechanical Trowel.

Pozzolan. Hydraulic cement obtained by grinding a mixture of fused natural material, such as volcanic ash or trass with hydrated lime; can be used to replace some of the portland cement to lower cost and in some cases improve the concrete; also spelled puzzolan.

PPG. Polished Plate Glass.

PPI. Plastic Pipe Institute.

PPM. Parts Per Million.

Pratt Truss. A standard peaked roof truss configuration where the vertical web members and the top chord are in compression while the sloping web members and the bottom chord are in tension.

Praxis. The practice of a skill, art, trade, or profession.

Precast. A concrete component or member cast and cured in other than its final position.

Precast Beam. A concrete horizontal structural member that is cast and cured in other than its final position, on- or off-site.

Precast Column. A concrete column that has been cast and cured in other than its final position.

Precast Concrete. Concrete parts that are cast on- or off-site and, after curing and hardening, are installed in their final position of use.

Precast Lintel. A horizontal concrete beam placed over an opening that has been cast and cured in other than its final position.

Precast Manhole. An underground enclosed structure that has been cast and cured in other than its final position, usually cast in sections.

Precast Panel. A concrete member, cast and cured in other than its final position and relatively thin with respect to other dimensions.

Precast Septic Tank. A sewage system underground tank that has been cast and cured in other than its final position.

Precast Slab. A flat, horizontal, molded layer of reinforced concrete, cast and cured in other than its final position, sometimes referred to as precast planks, often containing hollow cores.

Precast Specialty. Special shapes or ornamental objects of precast masonry or concrete.

Precast Terrazzo. Marble aggregate concrete that is cast and cured in other than its final position, and ground smooth for decorative purposes.

Precast Wall. A wall that has been cast and cured in other than its final position.

Precipitate. 1. A substance separated from a solution by chemical action or by application of heat or cold. 2. Insoluble compound formed when chlorine or alum added to pool water reacts with other chemicals or minerals.

Precipitation. Rain, hail, or snow falling to the ground.

Precoat. The initial coating of filter aid on the septum of a diatomaceous earth filter.

Preconstruction Jobsite Conference. A conference on the jobsite, before start of construction, attended by owners, architects, engineers, contractors, and subcontractors for the purpose of coordination and cooperation.

Predecorated Wallboard. A gypsumboard with a finished surface applied before installation; gypsumboard may be predecorated with paint, texture, vinyl film, or printed paper coverings in a variety of patterns, styles, and colors.

Prefabricated. Constructed and assembled in a workshop and later brought to the jobsite for incorporation into the building.

Prefabricated Lined Stack. Chimney or vent fabricated in a shop into sections with multiple components to provide thermal insulation and allow for low clearance to wooden or combustible elements.

Pre-fill. An application method used in the preparation of tapered/beveled gypsum board to receive tape and joint treatment; aids in reducing the possibility of ridging and beading.

Prefinished Gypsumboard. Gypsumboard finished at the factory with a decorative layer of paint, paper, or plastic.

Pre-Function Area. An anteroom in a hotel or conference center leading to a large dining room, used for receptions prior to dinners or meetings.

Preheating. The application of heat to the base metal prior to a welding or cutting operation.

Prehung Door. A door that is hinged to its frame in a factory or shop.

Preliminary Drawings. Drawings prepared in a preliminary stage of a project.

Preliminary Notice. A notice sent to the owner, general contractor, and lender by subcontractors or materialmen within 20 days of first supplying labor, services or materials to a project informing them of what is being supplied or performed and establishing lien or stop-notice rights.

Premature Stiffening. See False Set.

Pre-Mixed Plaster. See Gypsum Ready Mixed Plaster.

Prepacked Concrete. Concrete manufactured by placing clean, graded coarse aggregate in a form and later injecting a portland cement sand grout, under pressure, to fill the voids.

Prepaid Expenses. Materials or services that a company buys and pays for before use, such as insurance premiums, or office supplies.

Preparation, Painting. See Painting Preparation.

Prepayment Penalty. A monetary penalty for paying off a note prior to its due date.

Present Value. The value now of a sum of money to be paid or received in the future, calculated by discounting future cash flows by an appropriate discount rate.

Preservation. 1. The process of treating materials to prolong their useful life. 2. Conservation.

Preservative. 1. A substance for preserving. 2. In wood, any substance that, for a reasonable length of time, is effective in preventing the development and action of wood-rotting fungi, borers of various kinds, and harmful insects that deteriorate wood.

Preservative Treated. Applied or pressurized chemical treatment of wood or plywood to make it resistant to deterioration from moisture and insects.

Press. A machine for punching, notching, and shearing steel.

Pressed Bricks. Bricks that are pressed out in a mold before they are baked or burned.

Presser, Clothing. See Clothing Presser.

Pressure. Energy impact on a unit area; force or thrust on a surface.

Pressure, Absolute. See Absolute Pressure.

Pressure, Atmospheric. See Atmospheric Pressure.

Pressure, Back. See Back Pressure.

Pressure Balance. In spray painting, relationship of pot pressure to atomizing air pressure.

Pressure Connector. An electrical connecting device which squeezes two or more conductors together.

Pressure Connector, Solderless. See Solderless Pressure Connector.

Pressure Differential. Pressure difference across hydraulic system.

Pressure Drop. 1. Loss in pressure due usually to length or diameter of line or hose. 2. Pressure difference at two ends of a circuit, or part of a circuit, the two sides of a filter.

Pressure Feed Paint Tank. In spray painting equipment, fluid container in which fluid flow is caused by air pressure; also called Pressure Pot.

Pressure Feed. In spray painting equipment, fluid flow caused by application of air or hydraulic pressure to paint.

Pressure-Heat Diagram. Graph of refrigerant pressure, heat, and temperature properties; see Mollier's diagram.

Pressure Gauge. 1. A gauge for measuring the pressure exerted by a liquid or a gas. 2. A device to measure the pressure of an explosive.

Pressure, Gauge. See Gauge Pressure.

Pressure Head. Force caused by the weight of a column or body of fluids; expressed in feet, inches or psi.

Pressure Head. The amount of force created by a depth of one foot of water.

Pressure Limiter. Device which remains closed until a certain pressure is reached and then opens and releases fluid to another part of system or breaks an electric circuit.

Pressure Motor Control. High or low-pressure control connected into the electrical circuit and used to start and stop motor; it is activated by demand for refrigeration or for safety.

Pressure-Operated Altitude Valve (POA Valve). Device which maintains a constant low-side pressure independent of altitude of operation.

Pressure, Operating. Pressure at which a system is operating.

Pressure Pot. See Pressure Feed Paint Tank.

Pressure Reducing Valve. A valve which maintains fluid pressure uniformly on its outlet side as long as pressure on the inlet side is at or above a design pressure.

Pressure Regulator, Evaporator. See Evaporator Pressure Regulator.

Pressure Regulator. Automatic valve located between a compressor and evaporator outlet that is responsive to its own inlet pressure or to the evaporator or refrigerator temperature; it throttles the vapor flow when necessary to prevent the evaporator pressure from falling below a selected value.

Pressure Regulator. A valve which controls water pressure in the supply line.

Pressure-Sensitive Adhesive. An adhesive that will adhere to a surface at room temperature by briefly applied pressure alone.

Pressure, Suction. Pressure in low-pressure side of a refrigerating system.

Pressure Switch. An electric switch that is activated by a rise or drop in air or fluid pressure.

Pressure-Treated Lumber. Lumber that has been impregnated with chemicals under pressure, for the purpose of retarding either decay or fire.

Pressure Water Valve. Device used to control water flow; it is responsive to head pressure of refrigerating system.

Prestressed Beam. A horizontal structural member which has had a load applied to it to increase its effectiveness in resisting working loads.

Prestressed Concrete. Concrete in which the steel is stretched and anchored to compress the concrete.

Prestressing. Applying compressive stress to a concrete structural member to increase its strength.

Pretax Income. Net income or profit shown on income and expense statement before deduction for Federal Taxes.

Pretensioning. A method by which the design tensile force is applied to the steel reinforcing before the concrete is set.

Pretreatment. Chemical alteration of the surface to make it suitable for painting.

Prevailing Wage Law. A law that establishes minimum wages for job classifications in the construction trades.

Prevalent Level Samples. Air samples taken under normal conditions; also called Ambient Background Samples.

Preventive Maintenance Painting. Periodic touch-up painting or application of full coats of paint before deterioration starts.

Price. The amount of money that must be paid for something.

Primary Branch. This is a single branch that slopes from the base of a soil or waste stack and joins the main building drain or another branch of the main building drain.

Primary Color. One of the colors from which all other colors may be produced by mixing; in pigments, red, yellow, and blue.

Primary Control. Device which directly controls operation of heating system.

Prime Coat. 1.The first coating of any material placed on a surface to seal and to provide a proper base for additional finish coats. 2. The first coat of paint.

Prime Contractor. A general contractor who contracts with a property owner and, in turn, employs a subcontractor or subcontractors to perform some or all of the work.

Primed Door. A door which has received its first coat of paint prior to installation.

Primed Siding. Exterior siding material that has received its first coat of paint.

Prime In the Spots. Apply a priming coat to those spots that have been scraped, wire brushed, shellacked, have had the old paint burned off ,or consist of newly patched plaster.

Prime Professional. Any professional having a contract directly with the owner.

Primer. 1. A tinted base coat of paint to seal the surface and equalize suction differences. 2. A liquid composed of bituminous solvent that is applied to a surface to improve the adhesion of heavier subsequent applications of bitumen.

Prime Rate. The rate of interest paid by the most credit-worthy customers of a bank.

Primer Paint. A specially mixed paint that is applied first to a surface to seal and provide a base for additional finish coats.

Principal. 1. Under a surety bond, the contractor is the principal. 2. The amount of a loan or the loan balance at any particular time. 3. In a debt repayment, the principal portion goes to reducing the debt as opposed to the interest portion.

Principle. 1. A fundamental truth or law as the basis of reasoning or action. 2. A general law in physics, chemistry, or engineering. 3. A law of nature forming the basis for the construction or working of a machine or mechanical device.

Print Dyeing. Screen printing a pattern on carpet by successive applications of premetalized dyes, which are driven into the pile construction by an electromagnetic charge.

Print Free. Paint sufficiently dry so that no imprint is left when something is pressed against it.

Prism. 1. A solid geometric figure whose two ends are similar, equal, and parallel rectilinear figures, and whose sides are parallelograms. 2. A transparent body in this form, usually triangular, with refracting surfaces at an acute angle with each other, which separates white light into a spectrum of colors.

Prism Testing of Masonry. Compression testing of a sample section of wall approximately 16 inches high and 16 inches long of the thickness and type of construction similar to the wall under construction.

Privacy Lock. A lockset, without a key, that can be locked from the room side only; used mainly on bedroom and bathroom doors; also called a Bathroom Lock.

Private Mortgage Insurance. Life insurance carried by a home buyer to pay off the mortgage in the event of buyer's death; the insured amount declines as the loan is paid off; this type of insurance is usually required by the lender when the loan to value ratio exceeds 90%.

Private Sewer. A sewer privately owned.

Privity. Persons who have contractual relationships with each other are in privity.

Pro Forma. As a matter of form; provided in advance to prescribe form.

Pro Forma Balance Sheet. A statement representing a future financial position of a business showing anticipated assets, liabilities and net worth.

Pro Rata. Proportionate or pro-rated.

Probability. The extent to which an event is likely to occur, measured by the ratio of the favorable cases to the whole number of cases possible.

Probate. The official court process of proving the validity of a will.

Procedure. Rules governing court operations.

Process. 1. A procedure for manufacturing products. 2. To handle or deal with a procedure.

Processed Lime. Pulverized quick lime.

Processed Quicklime. See Lime.

Processor, Photo. See Photo Processor.

Process Tube. Length of tubing fastened to hermetic unit dome, used for servicing unit.

Produce Case. A bin or container, usually open, that holds market produce, such as lettuce or tomatoes.

Product Data. Illustrations, standard schedules, performance charts, instructions, brochures, diagrams, and other information furnished by the contractor to illustrate materials or equipment for some portion of the work.

Product Liability. A manufacturer's legal liability for claims arising out of damages sustained by users of the product.

Production Rate. Measurement of surface cleaned or coated in one working day by one person, usually in units of square feet per day.

Professional Engineer. A person qualified as an engineer and duly licensed by the state to practice as a civil, structural, electrical, mechanical, or other engineer.

Professional Fee. The amount of money charged by a person hired to perform a professional service.

Professional Indemnity Insurance. See Errors and Omissions Insurance.

Professional Liability. A professional's legal liability for claims arising out of damages sustained by others allegedly as a result of negligent acts, errors, or omissions in the performance of professional services.

Professional Standard of Care. The duty of an architect, engineer, or other professional to exercise such care, skill, and diligence as other professionals in the same discipline would in the same or similar circumstances.

Profile. 1. The outline of something. 2. A drawing of something in section. 3. Surface contour of a blast-cleaned surface as viewed from the edge.

Profile Depth. Average distance between top of peaks and bottom of valleys on the surface.

Profit. The excess of income over expenditure; net income.

Profit and Loss. A summary account used at the end of an accounting period in which income and expenditure and the resulting net profit or loss are shown.

Profit and Loss Statement. A financial statement showing the income and expenses of a business for the period, and profit or loss resulting therefrom.

Program. A detailed outline of all of the features to be included in a building.

Program Evaluation Review Technique (PERT). An early form of critical path analysis for construction scheduling.

Progressive Blocking. Wood blocking between joists or rafters progressively nailed as each joist or rafter is installed.

Progressive Kiln. A kiln in which the total charge of lumber is not dried as a single unit but as several units, such as kiln truckloads, that move progressively through the kiln; the kiln is designed so that the temperature is lower and the relative humidity higher at the end where the lumber enters than at the discharge end.

Progressive Loss. An injury to person or property that occurs over a period of time, such as cracking caused by gradual settlement of soils or illness caused by inhalation of dust.

Project. 1. A scheme or design for an undertaking or development of real property. 2. To protrude or extend outward.

Project Delivery System. Method by which owners can contract for construction, such as design-bid-build, design/build, turnkey, or other of numerous variants.

Projecting Sash. A window with sashes that swing either inward or outward.

Projecting Window. A window that swings either inward or outward.

Project Manual. The volume usually assembled for the work which may include the bidding requirements, sample forms, conditions of the contract, and specifications.

Projector, Nonprofessional. The nonprofessional projector employs film other than 7/8 inch wide.

Projector, Professional. The professional projector employs a film which is more than 7/8 inch wide.

Promise. An undertaking that something will or will not happen in the future.

Promisee. A person to whom a promise is made.

Promisor. A person who makes a promise.

Promissory Estoppel. A promise that is enforceable even though it is not supported by consideration, since the promisor is estopped to withdraw it.

Promissory Note. A written promise to pay a sum on or before a particular date.

Proof of Service. Evidence that a legal document has been delivered to a specified person.

Propane. 1. A heavy gaseous hydrocarbon of the alkane series used as bottled fuel. 2. Low temperature application refrigerant.

Propeller Mixer. A type of mixing device which is inserted into a liquid, usually paint, and powered by a propeller at the end of a thin shaft.

Property. A tangible thing that is, or may be, possessed by a person.

Property Damage. Injury to tangible property.

Property Insurance. Insurance against damage to, or loss of, property that belongs to the insured.

Proportional Limit. The maximum unit stress for which Hooke's Law is valid.

Proportioning. Selection of proportions of material for concrete to make the most economical use of available materials to manufacture concrete of the required strength, placeability, and durability.

Proposal. The document submitted by the contractor to the owner for construction of the project; also called the Bid.

Propylaeum. (Pl. propylaea) An architecturally important vestibule or entrance to a building or enclosure; usually used in the plural.

Propylene Plastic. Plastics based on resins made by the polymerization of propylene with one or more other unsaturated compounds, the propylene being in greatest amount by weight.

Proration. The pro rata distribution of expenses in proportion to time of transaction, as taxes and insurance in a property sale escrow or as in shared expenses in proportion to space leased.

Proscenium Arch. The wall that separates the stage from the auditorium and provides the arch that frames it.

Protected Membrane Roof. A membrane roof assembly in which the thermal insulation lies above the waterproof membrane.

Protective Board. A board or sheet of material that is installed next to a waterproofing membrane and then backfilled against thus protecting the membrane from puncture or damage.

Protective Life. Interval of time during which a paint system protects substrate from deterioration.

Protector, Circuit. See Circuit Protector.

Proton. Particle of an atom with a positive charge.

Prototype. 1. An operational model of a new product. 2. An original design that is the basis of all future improvements and variations; an Archetype.

Protractor. A device for measuring angles on drawings or in constructing angles on drawings.

Provisional Remedies. Orders or proceedings that protect persons or property while legal proceedings are pending.

Prussian Blue (Ferric Ferrocyanide). Pigment which is deep in color and has great strength. Not affected by acids; easily affected by alkali.

Prybar. An iron or steel bar used for prying, such as a crowbar.

PS. Products Standards Section, U.S. Dept. of Commerce.

PSF. Pounds per square foot.

PSI. Pounds per square inch.

PSIA. Pounds per Square Inch Absolute.

PSIG. Pounds per Square Inch Gauge.

Psychological Factors. Psychological, organizational, and personal stresses that could produce symptoms similar to poor indoor air quality.

Psychrometer. See Hygrometer.

Psychrometric Chart. Chart that shows relationship between the temperature, pressure, and moisture content of the air.

Psychrometric Measurement. Measurement of temperature, pressure, and humidity using a psychrometric chart.

Psychrometrics. A branch of physics dealing with the study of atmospheric conditions, specifically the relationships between moisture and air.

P-Trap. A P-shaped drain trap that prevents sewer gas from escaping from a plumbing fixture.

P-Trap, ABS. A P-trap used mostly at sinks and lavatories; manufactured of acrylonitrile butadiene styrene, a rigid plastic pipe.

P-Trap, No-Hub. A P-trap used mostly at sinks and lavatories; made to join with no-hub cast iron drain piping.

Publication. The process of making writings or drawings available for distribution to or inspection by the public.

Public Sewer. A sewer that is publicly owned.

Public Utility Regulatory Policies Act. A law requiring public utility companies to buy on-site generated electricity from private power producers.

Public Works. Construction and projects undertaken on behalf of governmental entities.

Puckering. A condition in a carpet seam, due to poor layout or unequal stretching, wherein the carpet on one side of the seam is longer or shorter than that on the other side, causing the long side to wrinkle or develop a pleated effect.

PUD. Planned Unit Development.

Puddle. To compact loose soil by soaking it and allowing it to dry.

Puddle Welded. Spot weld, made by arc welding process, in which coalescence proceeds from surface of one member into other; weld is made without preparing a hole in either member.

Puddling. A condition of mechanical dash textures resulting in glazing, texture deviation or discoloration caused by holding the plastering machine nozzle too long in one area.

Puffing Agent. A synthetic organic product used to produce increased viscosity in varnishes and paints.

Pull Box. An electrical rough-in box placed in a length of conduit, through which cables can be pulled.

Pull Chain. An electric switch at a fixture operated by a pull cord.

Pulley. A grooved wheel for a cord, chain, or belt to pass over, used for changing the direction of a force or for mechanical advantage.

Pulley Stile. The members of a window frame which contain the counterweights and pulleys, and between which the edges of the sash slide.

Pulling. See Floating.

Pull Saw. A handsaw that cuts on the pull rather than on the push stroke.

Pull Wire. A wire installed in a conduit by an electrician by which additional wires are pulled through.

Pulp. Wood fiber added as an aggregate to neat calcined gypsum.

Pulvinar. A cushioned seat.

Pumice. A light porous volcanic rock often used as an abrasive and as plaster aggregate.

Pumice Stone. A stone of volcanic origin, which is pulverized to produce a soft abrasive used extensively in rubbing finishing coats of fine wood finishes.

Pump. 1. A machine usually with a rotary action or the reciprocal action of a piston, for raising or moving liquids, compressing gases, spray painting, and inflating tires. 2. See Plastering Machine.

Pump, Centrifugal. See Centrifugal Pump.

Pump Down. The act of using a compressor or a pump to reduce the pressure in a container or a system.

Pumped Concrete. Concrete that is pumped through a hose or pipe.

Pump, Fixed Displacement. See Fixed Displacement Pump.

Pump, Gas. See Gas Pump.

Pumping Agent. A product used to increase the flow of plaster through hoses during machine applications.

Pumping of a Slab. The ejection of mud and water up through joints or cracks when a vehicle passes over an on-grade concrete slab.

Pump, Reciprocating Single Piston. See Reciprocating Single Piston Pump.

Pump, Screw. See Screw Pump.

Pump, Sump. See Sump Pump.

Punch. A small hand tool formed from a short steel bar with one end in various shapes to suit the work, such as forming, perforating, embossing, or cutting, when struck by a hammer.

Punched Carpet. Carpet made by punching loose, unspun fibers through a woven sheet which results in a pileless carpet similar to a heavy felt; usually consists entirely of synthetic fibers.

Punch List. See Inspection List.

Punch Out. A close fitting hole in gypsum board to allow penetration of plumbing lines; a pipe hole.

Punitive Damages. Damages awarded to a private person against a wrongdoer by way of punishment, and to deter future misconduct.

Punkiness Soft or spongy gypsumboard core.

Purchase Money Mortgage. A mortgage or trust deed given to a seller as part of the purchase consideration.

Purchases Account. The account in which merchandise purchased for resale is recorded.

Purger, Air. See Air Purger.

Purging. Releasing compressed gas to atmosphere through some part or parts for the purpose of removing contaminants from that part or parts.

Purlin. 1. A timber supporting several rafters at one or more points. 2. Beams or struts that span across a roof to support the roof framing system.

Push Plate. A metal plate on a door installed at hand level for pushing the door open, used for protection from damage and easier cleaning.

Push Stick. A pole or strip used to push a workpiece when cutting with power saws, jointers and other power tools; pushing a board by hand is unsafe with power equipment.

Putty. A cement composed of whiting and raw linseed oil, used for setting panes of glass and filling small holes and cracks.

Putty Coat. Final smooth coat of plaster.

Putty Knife. A broad-bladed hand tool which is used to dispense and apply putty to surfaces.

PV. Photovoltaic.

PVA. Polyvinyl Acetate.

PVB. Polyvinyl Butyral.

PVC. 1. Polyvinyl Chloride. 2. Pigment Volume Concentration.

PVC Caulking. Caulking made of polyvinyl chloride.

PVC Cement. A solvent cement specifically used to join PVC pipe and fittings.

PVC Chamfer Strip. A triangular or curved insert constructed of PVC placed in an inside form corner to produce a rounded or flat, beveled edge.

PVC Conduit. Lengths of rigid plastic pipe made of polyvinyl chloride.

PVC Control Joint. Control joint made of polyvinyl chloride.

PV Cells. Photovoltaic cells.

PVC Fitting. A fitting made of polyvinyl chloride used in joining PVC piping.

PVC Pipe. Polyvinyl chloride pipe used mainly for drain lines and particularly resistant to chemicals.

PVC Sewer Pipe. Large diameter pipe used for the dispersal of waste material, constructed of polyvinyl chloride, a rigid plastic pipe.

PVC Waterstop. A waterstop made of polyvinyl chloride.

P-Wave. The primary or fastest wave traveling away from a seismic event through the earth's crust, and consisting of a train of compressions and dilations of the material; see Seismic Wave.

Pyramid. A polyhedron having for its base a polygon and for its faces triangles, meeting at a common apex.

Pyroligneous. Obtained by destructive distillation of wood.

Pyrolysis. Chemical change brought about by the action of heat.

Pyrometer. A thermometer for measuring very high temperatures.

Pyroxylin. See Cellulose Nitrate.

Pythagorean Theorem. The square of the length of the hypotenuse of a right-angled triangle is equal to the sum of the squares of the lengths of the other two sides.